Katrin Niedermann

Direct N-Trifluoromethylation

Katrin Niedermann

Direct N-Trifluoromethylation

Direct Trifluoromethylation of Organonitrogen
Compounds with Hypervalent Iodine Reagents

Südwestdeutscher Verlag für Hochschulschriften

Impressum / Imprint

Bibliografische Information der Deutschen Nationalbibliothek: Die Deutsche Nationalbibliothek verzeichnet diese Publikation in der Deutschen Nationalbibliografie; detaillierte bibliografische Daten sind im Internet über http://dnb.d-nb.de abrufbar.

Alle in diesem Buch genannten Marken und Produktnamen unterliegen warenzeichen-, marken- oder patentrechtlichem Schutz bzw. sind Warenzeichen oder eingetragene Warenzeichen der jeweiligen Inhaber. Die Wiedergabe von Marken, Produktnamen, Gebrauchsnamen, Handelsnamen, Warenbezeichnungen u.s.w. in diesem Werk berechtigt auch ohne besondere Kennzeichnung nicht zu der Annahme, dass solche Namen im Sinne der Warenzeichen- und Markenschutzgesetzgebung als frei zu betrachten wären und daher von jedermann benutzt werden dürften.

Bibliographic information published by the Deutsche Nationalbibliothek: The Deutsche Nationalbibliothek lists this publication in the Deutsche Nationalbibliografie; detailed bibliographic data are available in the Internet at http://dnb.d-nb.de.

Any brand names and product names mentioned in this book are subject to trademark, brand or patent protection and are trademarks or registered trademarks of their respective holders. The use of brand names, product names, common names, trade names, product descriptions etc. even without a particular marking in this works is in no way to be construed to mean that such names may be regarded as unrestricted in respect of trademark and brand protection legislation and could thus be used by anyone.

Coverbild / Cover image: www.ingimage.com

Verlag / Publisher:
Südwestdeutscher Verlag für Hochschulschriften
ist ein Imprint der / is a trademark of
AV Akademikerverlag GmbH & Co. KG
Heinrich-Böcking-Str. 6-8, 66121 Saarbrücken, Deutschland / Germany
Email: info@svh-verlag.de

Herstellung: siehe letzte Seite /
Printed at: see last page
ISBN: 978-3-8381-3620-2

Zugl. / Approved by: Zürich, ETH, Diss., 2012

Copyright © 2013 AV Akademikerverlag GmbH & Co. KG
Alle Rechte vorbehalten. / All rights reserved. Saarbrücken 2013

Ohne Flachs!
deutsche Redewendung

Für meine Familie

This dissertation was submitted to ETH Zurich and the original work is available via http://dx.doi.org/10.3929/ethz-a-007567196 at the ETH e-collection database.

Acknowledgement

During my studies, I have been supported by several people. Without their help, profound knowledge and their willingness to guide and support me, I would not have been able to complete the work presented in this thesis. Therefore I would like to express my gratitude to the following people and institutions:

Prof. Dr. Antonio Togni, meinem Doktorvater, danke ich für die Chance, dass ich meine Doktorarbeit in seiner Arbeitsgruppe ausführen durfte sowie sein Vertrauen in meine Arbeit und die damit verbunden Freiheiten. Dass ich einen Teil meiner Arbeit ausserhalb der ETH, in der Arbeitsgruppe von *Prof. Dr. Konrad Seppelt* an der FU Berlin, durchführen konnte, um einige gewagte und spekulative Konzepte zu testen, hat mich sehr gefreut. Auch dafür herzlichen Dank. Deine Begeisterung für die Fluorchemie war ansteckend und sehr motivierend.

Prof. Dr. Konrad Seppelt möchte ich für die sehr lehrreiche Zeit an der FU Berlin sowie für die Übernahme des Korreferates danken. Die Chemie, wie sie in Ihrem Arbeitskreis ausgeführt wird, ist unvergleichlich. Ich bin sehr dankbar, dass ich die Möglichkeit hatte, diese zu erlernen.

Prof. Dr. Dieter Seebach danke ich für die Übernahme des Korreferates und die wohlwollende Unterstützung. Ausserdem möchte ich mich für die lehrreiche Kollaboration bezüglich der Trifluormethylierung von Sandostatin® bedanken.

SNF and *ETH Zurich* are acknowledged for funding. Also, I am grateful for having received the

SCNAT/SCS Travel award to finance my trip to the 242nd ACS national meeting.

Dr. Patrick Eisenberger und *Dr. Iris Kieltsch*, meinen Mentoren während meines Chemiestudiums, gilt ein besonderer Dank. Sie haben mir ihr reichhaltiges Wissen und die praktischen Kniffe der synthetischen Chemie weitergeben. In diesem Zusammenhang möchte ich mich auch bei

Dr. Raffael Koller für die Unterstützung und die unzähligen Ratschläge bedanken. Besonders für die Hilfestellung zu Beginn meiner Doktorarbeit bin ich sehr dankbar. Überdies hat seine Dissertation über die Aktivierung des Reagenz massgeblich zum Erfolg meiner Arbeit beigetragen.

Ein ganz grosser Dank gilt natürlich allen Kollegen, die auf diesem Gebiet Vorarbeit geleistet haben, insbesondere der Pionierarbeit von *Patrick* und *Iris*, sowie die mechanistischen Untersuchungen von

Dr. Jan M. Welch. I am deeply grateful for his patient explanations of mechanistic concepts, his help in the fluorine business and his company at the meetings in Ljubljana and Denver. Furthermore, I am thankful for the careful proofreading of this manuscript.

Barbara Czarniecki gebührt ebenfalls ein grosser Dank. Auch sie hat die vorliegende Doktorarbeit korrigiert. Zusätzlich möchte ich mich für die von ihr oder von uns gemeinsam organisierten Gruppenausflüge bedanken. Ebenfalls werden ihr Besuch in Berlin und die „Tuesday evenings" in guter Erinnerung bleiben. Zusammen mit

Dr. René Verel, Dr. Aitor Moreno und *Dr. Heinz Rüegger* hat sie immer hilfreich alle Fragen rund ums NMR beantwortet.

Dr. Ján Cvengroš, Dr. Raffael Koller, Nico Santschi und *Philip Battaglia* danke ich für deren synthetischen Beitrag im Strukutur- und Reaktivitätsteil im Kapitel 2 dieser Dissertation sowie *Dr. Jan M. Welch* für seinen Beitrag zur Reaktivitätsstudie.

Ekaterina Vinogradova and *Dr. Matthias S. Wiehn*, I am thankful for the initial discovery of the Ritter-type reaction, without their initial experiments, the core of this dissertation wouldn't be as comprehensive. До скорой встречи в Бостоне.

Halua Pinto de Magalhães, Oliver Sala und *PD Dr. Hans Peter Lüthi* danke ich für die umfassenden Berechnungen zur Aufklärung des Reaktionsmechanismus der Ritter-artigen Reaktionen.

Dr. Engelbert Zass danke ich für die umfassende Literaturrechere über NCF_3 Verbindungen.

I want to thank all former and current members from the *Togni-* and *Mezzetti-group* for the great time in, around, and outside the lab during the last four years. Gebührend bedanken möchte ich mich bei meinen Kollegen aus dem H222. Die aufgeschlossene und angenehme Arbeitsatmosphäre ist Teil meines Erfolgs. Insbesondere ist

Dr. Jonas Bürgler zu erwähnen. Wir haben die längste Zeit zusammen in diesem Labor gearbeitet und Jonas ist mir bei vielen kleineren und grösseren Problemen mit Rat und Tat zur Seite gestanden. Zusätzlich danke ich ihm für unsere kleine Kollaboration im „InnoCentive Projekt".

Raphael Aardocm danke ich für die puffernde Wirkung, egal ob im Labor oder in der WG. Ausserdem möchte ich nicht vergessen, dir für die vielen Ratschläge und Gefälligkeiten zu danken.

Remo Senn danke ich für die Zusammenarbeit bei der direkten Stickstoff Trifluormethylierung und wünsche ihm alles Gute für die anschliessenden Arbeiten.

Jolanda Winkler, unserer Lehrtochter, danke ich für ihre gewissenhafte Arbeit und die Freude an ihrer Tätigkeit.

Philip Battaglia, unserem ehemaligen Chemielaboranten Lehrling, danke ich für die Übernahme der einen oder anderen Synthese sowie das stets aufgeräumte Labor am Freitagabend.

Natalja Früh gilt ein besonderer Dank für ihren enthusiastischen Einsatz und ihre Zielstrebigkeit während ihrer Bachelor- and Semesterarbeit.

Matthias Geibel und *Rebekka Schwaninger* danke ich für ihren Einsatz während ihrer Semesterarbeiten. Besonders erwähnen möchte ich ihren Willen zur selbständigen Arbeit.

Raphael Aardoom, *Dr. Matthias Vogt*, *Amos Rosenthal*, *Dr. Pietro Butti*, *Dr. Francesco Camponovo*, *Dr. Michael Wörle*, *Prof. Dr. Antonio Mezzetti* und *Dr. Bernd Schweizer* möchte ich für die angenehme Zusammenarbeit, Hilfe und lehrreichen Diskussionen bezüglich Kristallographie danken.

Peter Ludwig und *Raphaël Rochat* danke ich für die Engelsgeduld und die Lösung all meiner kleineren und grösseren Computerprobleme.

Bei den *Arbeitskreisen Seppelt* und *Lentz* in Berlin möchte ich mich herzlich für eine unvergessliche Zeit sowie die hilfreichen Tipps bedanken; Chemie wird eben nicht überall gleich ausgeübt. Vielen Dank.

Rita Friese danke ich für die Messungen der Ramanspektren.

Danke an alle *Mitarbeiter des MS-* und *EA-Services* für ihre Messungen, auch wenn meine Proben sehr flüchtig waren oder etwas viel Fluor enthielten. Ebenso möchte ich mich bei den Technikern aus der *Zentralwerkstatt* für ihre kompetente Hilfe bei nicht alltäglichen Wünschen bedanken.

Vielen Dank all *meinen Freunden*, die mich den Stress und Druck bei einem Essen, beim Reiten, beim Sport oder „eis Trinka" vergessen liessen und mir immer wieder gezeigt haben, dass es auch eine Welt – eine durchaus schöne – ausserhalb der ETH gibt.

Marius Mewald danke ich für die herzliche Unterstützung, sein Verständnis und seine Geduld. Lieber Dank, dass du mir während den letzten vier Jahren immer wieder Rückhalt, Zuversicht und Vertrauen gegeben hast trotz der grossen örtlichen Distanz.

Meiner Familie gilt abschliessend ein besonderer Dank. Meinen Eltern *Ruth* und *Markus* bin sehr dankbar für ihre stetige und grossartige Unterstützung, sowie ihren Rückhalt während meiner ganzen Studienzeit. Herzlichen Dank meinen Geschwistern *Sara*, *Andreas* und *Ursula*, die mir zur Seite standen und mich motiviert haben.

List of Publications

Part of the work described in this thesis has been published:

K. Niedermann, J. M. Welch, R. Koller, J. Cvengroš, N. Santschi, P. Battaglia, A. Togni, "New Hypervalent Iodine Reagents for Electrophilic Trifluoromethylation and their Precursors: Synthesis, Structure, and Reactivity" *Tetrahedron*, **2010**, *66*, 5753-5761. (Chapter 2)

K. Niedermann, N. Früh, E. Vinogradova, M. S. Wiehn, A. Moreno, A .Togni "A Ritter-type Teaction: Direct Electrophilic Trifluoromethylation at Nitrogen Atoms Using Hypervalent Iodine Teagents" *Angew. Chem. Int. Ed.* **2011**, *50* (5), 1059-1063. (Section 3.2)

K. Niedermann, N. Früh, R. Senn, B. Czarniecki, R. Verel, A. Togni "Direct Electrophilic N-Trifluoromethylation of Azoles by a Hypervalent Iodine Reagent" *Angew. Chem. Int. Ed.* **2012**, *51*, 6511-6515. (Section 3.4)

Crystallographic contributions to further publications:

J. F. Buergler, K. Niedermann, A. Togni "P-Stereogenic Trifluoromethyl Derivatives of Josiphos: Synthesis, Coordination Properties and Applications in Asymmetric Catalysis" *Chem. Eur. J.* **2012**, *18*, 632-640.

R. Koller, K. Stanek, D. Stolz, R. Aardoom, K. Niedermann, A. Togni "Zinc-mediated Formation of Trifluoromethyl Ethers from Alcohols and Hypervalent Iodine Trifluoromethylation Reagents" *Angew. Chem. Int. Ed.* **2009**, *48*, 4332-4336.

Contributions in form of poster presentations at international conferences:

"Towards a New Class of Cationic Hypervalent Iodine Reagents for Electrophilic Trifluoromethylation" K. Niedermann, J. M. Welch, J. Cvengros, N. Santschi, A. Togni; 19th International Symposium on Fluorine Chemistry (19th ISFC), Jackson Hole (USA), August 2009.

"New Hypervalent Iodine Reagents for Electrophilic Trifluoromethylation: Synthesis, Structure, and Reactivity" K. Niedermann, J. M. Welch, A. Togni; 16th European Symposium on Fluorine Chemistry (16th ESFC), Ljubljana (Slovenia), July 2010.

A contribution in form of an invited oral presentation entitled:

"Direct Electrophilic Trifluoromethylation of Nitrogen Centers" K. Niedermann, N. Früh, R. Senn, E. Vinogradova, M. S. Wiehn, A. Moreno, A. Togni; 242nd ACS National Meeting, Denver (USA), August 30, 2011.

Table of Contents

	Abstract	iii
	Zusammenfassung	v
1	**Introduction**	**1**
1.1	General Aspects	1
1.2	Functional Group Interconversion	2
1.2.1	Deoxofluorination	2
1.2.2	Oxidative Desulfurization Fluorination	3
1.3	Direct Methods	3
1.3.1	Nucleophilic Trifluoromethylation	3
1.3.2	Radical Trifluoromethylation	6
1.3.3	Electrophilic Trifluoromethylation	6
2	**Structure and Reactivity Correlation of Hypervalent Iodine Reagents for Electrophilic Trifluoromethylation**	**11**
2.1	Introduction	11
2.2	Synthesis of Derivatives	13
2.2.1	Synthesis of 1-Chloro-λ^3-iodanes	13
2.2.2	Synthesis of hypervalent Trifluoromethyl Compounds	16
2.3	Solid State Structure Analysis	19
2.3.1	X-Ray Structures of 1-Chloro-λ^3-iodanes	19
2.3.2	X-Ray Structures of 1-(Trifluoromethyl)-λ^3-iodanes	23
2.4	Reactivity Study	25
2.5	Conclusion and Outlook	26
3	**Direct Trifluoromethylation of Organonitrogen Compounds**	**27**
3.1	Introduction	27
3.2	A Ritter-Type Reaction	28
3.2.1	Results	28
3.2.2	Structure Determination	32
3.3	Mechanistic Investigations	34

3.4	Direct *N*-Trifluoromethylation of Benzophenone Imine	38
3.5	Direct *N*-Trifluoromethylation of Heterocycles	39
3.5.1	Reaction Optimization	39
3.5.2	Substrate Scope	43
3.5.3	Product Characterization	46
3.5.4	Ongoing Work	47
3.6	Conclusion and Outlook	50
4	**General Conclusion, Comments and Outlook**	**53**
5	**Experimental Part**	**55**
5.1	General Remarks	55
5.1.1	Techniques	55
5.1.2	Analytical Methods	55
5.1.3	Chemicals	57
5.2	Hypervalent Iodine Compounds	58
5.3	Nitrogen-Center Nucleophiles	69
5.3.1	Ritter-type Reaction	69
5.3.2	Direct *N*-Trifluoromethylation	77
5.4	Rate Study	90
5.4.1	Trifluoromethylation of *para*-Toluenesulfonic Acid Monohydrate	90
5.4.2	Formation of (*E*)-*N*-(1-(1H-Benzo[*d*][1,2,3]triazol-1-yl)ethyliden)trifluoromethanamine (39)	91
5.4.3	*N*-Trifluoromethylation of 1-(Trimethylsilyl)-1*H*-benzo[*d*]triazole (65)	92
6	**Literature**	**95**
7	**Appendix**	**103**
7.1	Abbreviations	103
7.2	Crystallographic Data	105

Abstract

The synthesis, characterization and reactivity of electrophilic trifluoromethylating agents based on hypervalent iodine(III) compounds is the topic of the first part of this thesis (Chapter 2). For a comparative structural study several new five- and a six-membered heterocyclic 1-chloro-λ^3-iodanes, including two cationic species, were prepared. Three of them were further converted to the corresponding trifluoromethylating agents and their X-ray structures where compared with those of the respective 1-chloro derivatives. A reactivity study on the trifluoromethylation of *para*-toluenesulfonic acid was conducted in order to compare initial rates. Although this study should be taken as a qualitative guide, it can be concluded that compounds having a weakend I–O bond display a higher reactivity.

The second part of this thesis (Chapter 3) describes two methods for the *N*-trifluoromethylation of organonitrogen compounds utilizing the aforementioned hypervalent iodine compounds. Firstly, in a Ritter-type reaction, *N*-(trifluoromethyl)imine derivatives were prepared *via* acid-catalyzed trifluoromethylation of nitriles in the presence of various azoles.

In general, moderate to good yields are obtained for acetonitrile and propionitrile, while for more sterically demanding or conjugated nitriles such as iPrCN, PhCH$_2$CN or PhCN significantly decreased yields are observed as compared to CH$_3$CN. However, the reaction is limited to azoles. If the reaction is carried out with benzophenone imine, unstable direct N-trifluoromethylated imine is observed in low yield. Moderate yields are obtained when one equivalent of bulky (TMS)$_3$SiCl instead of the acid catalyst is added to the reaction mixture. Furthermore, based on preliminary kinetic experiments as well as computational studies (in collaboration with PD H. P. Lüthi) a reaction mechanism for the Ritter-type reaction is proposed.

Secondly, the efficient and mild direct N-trifluoromethylation of various electron-rich heterocycles such as pyrazoles, indazoles, triazoles, tetrazoles and to a certain extent benzimidazole is described. To avoid the formation of Ritter-type products chlorinated solvents were used. In addition to reagent activation by a Brønsted acid, the substrate is activated by silylation as well. *In situ* silylation of the substrates by 1,1,1,3,3,3-hexamethyldisilazane (HMDS) in the presence of catalytic silica sulfuric acid (SSA), allows the synthesis of the desired N-trifluoromethylated products without the isolation of the silylated intermediates. Moderate to excellent yields for the direct N-trifluoromethylation of azoles are obtained when the reactions are carried out at high concentrations at 35 °C in the presence of catalytic amounts of HNTf$_2$ and LiNTf$_2$.

Under the above described conditions 14 different azoles were successfully N-trifluoromethylated. Substrates with alkyl, aryl and alkoxycarbonyl substituents undergo the desired reaction and various substitution patterns are tolerated. Typically, when working with unsymmetrically substituted substrates, isomeric product mixtures are obtained, that can be separated by flash column chromatography.

Both methods, the Ritter-type and the direct N-trifluoromethylation, provide ready access to a wide variety of stable N-CF$_3$ compounds; rare substances which are otherwise very difficult to obtain.

Zusammenfassung

Im ersten Teil dieser Dissertation (Kapitel 2) wird die Synthese, Charakterisierung und Reaktivität elektrophiler Trifluormethylierungsreagenzien basierend auf hypervalentem Iod(III) beschrieben. Mehrere fünf- und sechs-gliedrige heterocyclische 1-Chlor-λ^3-iodane, inklusive zwei kationische Vertreter, wurden für eine vergleichende Strukturuntersuchung hergestellt. Drei dieser Verbindungen wurden zu den entsprechenden Trifluormethylierungsreagenzien umgesetzt und die Kristallstrukturen wurden mit denen der 1-Chlorvorläufer verglichen. In einer Reaktionsstudie zur Trifluormethylierung von *para*-Toluolsulfonsäure wurden die Anfangsreaktionsgeschwindigkeiten der verschiedenen Reagensderivate miteinander verglichen. Obgleich diese Studie nur als Orientierungshilfe dienen sollte hat sich gezeigt, dass Verbindungen mit schwachen Sauerstoff-Iod Bindungen zu höherer Reaktivität neigen.

Im zweiten Teil dieser Dissertation (Kapitel 3) werden zwei Methoden zur *N*-Trifluormethylierung von Organostickstoffverbindungen beschrieben. Bei diesen elektrophilen *N*-Trifluormethylierungen wurden die oben erwähnten hypervalenten Iodverbindungen verwendet. Zuerst, wurden in einer Ritter-artigen Reaktion *N*-(Trifluormethyl)iminderivate hergestellt. Dabei wurden in einer säurekatalysierten Reaktion Nitrile in Gegenwart von Azolen trifluormethyliert.

Im Allgemeinen werden für diesen Reaktionstyp moderate bis gute Ausbeuten für Acetonitril und Propionitril erhalten. Sterisch anspruchsvollere und konjugierte Nitrile, wie zum Beispiel iPrCN, PhCH$_2$CN oder PhCN, führen im Vergleich zu CH$_3$CN zu signifikant schlechteren Ausbeuten. Dieser Reaktionstyp ist auf Azole beschränkt und so wird, wenn die Reaktion mit Benzophenonimin ausgeführt wird, instabiles direkt N-trifluormethyliertes Imin nur in geringer Ausbeute gebildet. Die Ausbeute kann verbessert werden, wenn anstelle des Säurekatalysators ein Äquivalent sterisch anspruchvolles (TMS)$_3$SiCl zur Reaktionsmischung gegeben wird. Zusätzlich, basierend auf ersten kinetischen Experimenten, sowie auf Computer gestützten Berechnungen (in Zusammenarbeit mit PD H. P. Lüthi) wird ein Reaktionsmechanismus für die Ritterartige Reaktion vorgeschlagen.

Im Anschluss wird die direkte N-Trifluormethylierung verschiedener elektronreicher Heterocyclen wie Pyrazolen, Indazolen, Triazolen, Tetrazolen und auch bis zu einem gewissen Grad Benzimidazolen unter milden Bedingungen beschrieben. Um die Bildung unerwünschter Produkte einer Ritterreaktion zu verhindern, werden chlorierte Lösungsmittel verwendet. Zusätzlich zur Aktivierung des Reagens durch die Zugabe einer Brønsted-Säure, wird das Substrat durch Silylierung aktiviert. Die luftempfindlichen silylierten Intermediate müssen dank einer *in situ* Synthesesequenz mittels 1,1,1,3,3,3,-Hexamethyldisilazan (HMDS) und einer katalytischen Menge auf Silika imobilisierte Schwefelsäure (SSA) nicht isoliert werden. Moderate bis exzellente Ausbeuten werden für die direkte N-Trifluormethylierung von Azolen erreicht, wenn die Reaktion in hochkonzentrierten Reaktionslösungen bei 35 °C und mit einer katalytischen Menge HNTf$_2$ und LiNTf$_2$ durchgeführt wird.

Unter den oben beschriebenen Bedingungen wurden 14 verschiedene Azole erfolgreich N-trifluormethyliert. Substrate mit Alkyl-, Aryl- und Alkoxycarbonyl-Substitutenten konnten erfolgreich umgesetzt werden, zudem wird ein breites Substitutionsmuster toleriert. Unsymmetrische Sustrate liefern meist Isomerengemische; diese können jedoch mittels Säulenchromatographie getrennt werden.

Beide Methoden, die Ritter-artige sowie die direkte N-Trifluormethylierung liefern einen schnellen Zugang zu einer Vielfalt von stabilen N-CF$_3$ Verbindungen, welche mit anderen Methoden nur schwer zugänglich sind.

1 Introduction

1.1 General Aspects

Only very few fluoro-organic metabolites have been identified in the biosphere and only about a dozen compounds containing fluorine atom(s) have been found in nature,[1] possibly due to the very low concentration of fluoride in seawater (1.3 mg/L[2]). However, fluorine is the most abundant halogen in the earth's crust ($5.85 \cdot 10^2$ mg/kg[2]). This has allowed fluoro-organic chemistry to become one of the most rapidly developing areas of life sciences in the last 50 years.[3] It is estimated that nowadays, as many as 30 to 40% of agrochemicals and 20% of pharmaceuticals on the market contain fluorine.[4] Among fluorinated compounds, trifluoromethyl substituted molecules constitute a particular class and have found a large number of industrial applications ranging from dyes and polymers to pharmaceuticals and agrochemicals.[5] The CF_3 group is a strong σ- (-I_σ) and π-acceptor (-I_π) and in α,β-unsaturated systems negative hyperconjugation is observed, causing the electron density at the β-carbon to be decreased. In medicinal chemistry, the synthesis of fluorinated compounds derived from a lead structure is of great interest since metabolic stability, physiochemical properties, conformation and protein-ligand interactions of that structure can be tuned by fluorination.[6] The introduction of a CF_3, OCF_3 or SCF_3 group often results in an enhanced lipophilicity of the molecule, facilitating the uptake across membranes and thus leading, together with a decreased metabolic susceptibility, to an increased bioavailability, allowing a lower dose of the drug to be administered.

This chapter will describe the synthetic methods for the introduction of CF_3 functional groups and will give an overview on the methods available; an extensive discussion covering all strategies can be found in a review by McClinton,[5] as well as in the PhD thesis of P. Eisenberger.[7] The available methods to introduce the CF_3 moiety are subdivided into "direct" and "functional group interconversion" methods. The functional group interconversion represents an indirect method, whereby a suitable precursor is transformed into the corresponding CF_3 group by fluorination. On the other hand, in the direct method the CF_3 core is delivered as a fully assembled entity. The direct method can be further subdivided into nucleophilic, free radical and electrophilic trifluoromethylation. Unfortunately, the later two cases are sometimes hard to distinguish since the products of these reactions can be of the same constitution.

Introduction

1.2 Functional Group Interconversion

In the classical variant of functional group interconversion, a suitable functional group (R-CCl$_3$, R-CS$_2$H or R-CO$_2$H) is transformed into the corresponding trifluoromethylated compound by highly reactive fluorine sources, e. g. elemental fluorine, HF, SF$_4$ etc.[5, 7] The handling of these hazardous and toxic reagents requires special equipment, know-how and safety precautions. Furthermore, the harsh reaction conditions necessary are tolerated by only a limited number of functional groups and therefore these methods are mainly used in industrial processes for the synthesis of starting materials or fluorinated building blocks. Two relatively moderate methods of functional group interconversion for the formation of a CF$_3$ group are briefly described in the following subchapters.

1.2.1 Deoxofluorination

In 1960 sulfur tetrafluoride (SF$_4$) was used successfully for the deoxofluorination of aldehydes, ketones, and carboxylic acids, to yield the corresponding -CF$_2$H, -CF$_2$- and -CF$_3$ compounds, respectively.[8] To circumvent the use of this highly toxic gas, diethylaminosulfur trifluoride (DAST) was developed as an alternative. Later, the thermally more stable analogue bis(2-methoxyethyl)aminosulfur trifluoride (Deoxo-Fluor™) was developed, since the larger scale application of DAST was limited by its well-known thermal instability. [9] The reactivity of these dialkylaminosulfur trifluorides closely mirrors the reactivity of their parent compound SF$_4$, but the only conversion of a benzoic acid into the corresponding trifluoromethylated compound has been reported for the parent compound yielding (trifluoromethyl)benzene (with DAST).[10] However, acyl fluorides react with Deoxo-Fluor™ to the corresponding trifluorides in moderate to good yields.[9] Recently, the synthesis, properties and reactivity of phenylsulfur trifluorides has been reported, and 4-*tert*-butyl-2,6-dimethylphenylsulfur trifluoride (Fluolead™) which has superior utility as a deoxofluorinating agent compared to current reagents was presented.[11] As shown in Scheme 1 a range of acids can be converted to the corresponding trifluoromethylated compounds when Fluolead is used as deoxofluorinating agent at elevated temperatures.

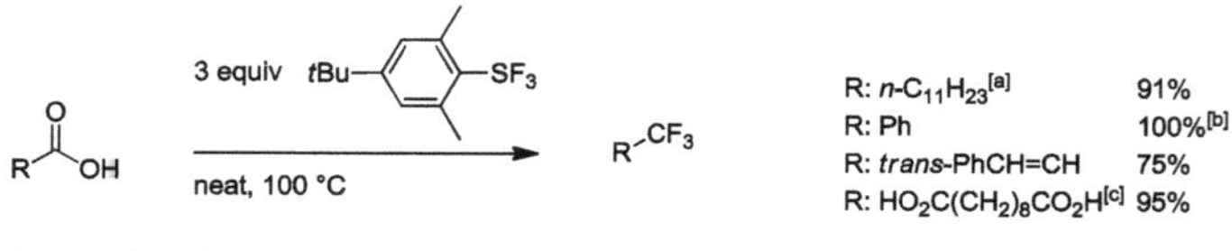

Scheme 1. Deoxofluorination of acids using Fluolead™ as fluorinating agent. [a] additive: 2.9 equiv HF-py, 50 °C; [b] yield determined by ^{19}F NMR; [c] 6 equiv of Fluolead™.[11]

1.2.2 Oxidative Desulfurization Fluorination

This method utilizes milder fluorinating agents in combination with N-haloimide oxidants to transform a methyl dithiocarboxylate into a CF_3 group.[12] HF-pyridine (Olah's reagent) or tetrabutylammonium dihydrogentrifluoride ($TBAH_2F_3$) can be utilized as fluoride sources for this type of reaction. The weak nucleophilicity of these reagents is compensated by the activation of the leaving group in the substrate by an electrophilic oxidation agent such as NBS, NIS or DBH (1,3-dibromo-5,5-dimethylhydantoin). As shown in Scheme 2, this method not only allows the synthesis of trifluoromethyl substituted (hetero)arenes under relatively mild conditions, but also the synthesis of trifluoromethyl ethers and N-trifluoromethylanilines.

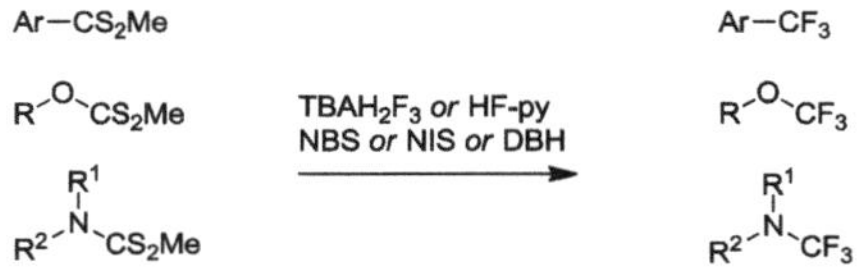

Scheme 2. Synthesis of trifluoromethylated (hetero)arenes, amines and ethers by oxidative desulfurization fluorination.[12c]

Instead of the above described fluorinating agents, BrF_3 can be utilized as well. The use of this reagent in organic chemistry has been described by Rozen.[13] The inherent oxidazing power of this nucleophilic fluorine source allows the conversion of a methyl dithiocarboxylate into a CF_3 group without an additional oxidant. Good yields are generally obtained for the conversion to the corresponding trifluoromethylated arenes, ethers and amines, but it should be noted that BrF_3 is a very corrosive material and should be used in well ventilated areas and in the absence of oxygenated solvents.

1.3 Direct Methods

1.3.1 Nucleophilic Trifluoromethylation

Among the strategies for the direct introduction of a trifluoromethyl group into organic molecules, nucleophilic trifluoromethylation has been the most attractive approach during the last few decades.[14] Historically, one of the first concepts accessed the use of MCF_3-type reagents. While the lithium and magnesium analogs show low thermal stability, late transition metals (Cu, Zn, Cd, Hg) and main group elements (Sn, Bi) are more suitable to stabilize the trifluoromethyl group by partial delocalization of the negative charge over low lying unoccupied metal orbitals. These reagents are normally applied in aromatic substitution reactions by thermal activation of aryl bromides or iodides. However, they suffer from low efficiency and normally yield various fluorinated side products. In 2008 Vicic reported the synthesis and isolation of the first thermally stable and well defined LCu(I)-CF$_3$ complex. In situ formed NHC-Cu(I)CF$_3$ complexes lead to good yields in a mild trifluoromethylation of organic halides at room temperature.[15] This work opened the door for the very recent advances in the metal (Cu and Pd) mediated or catalyzed trifluoromethylation of aryl halides,[16] aryl and heteroaryl boronic acids,[17] heteroarenes,[18] indoles,[19] vinyl sulfonates,[20] activated alkenes,[21] and terminal alkynes[22]. These transformations all utilize the Ruppert-Prakash reagent (Me$_3$SiCF$_3$), or the triethyl derivative, Et$_3$SiCF$_3$. This type of reagent led to the wide-ranging development of nucleophilic trifluoromethylation thanks to its easy handling and broad range of application. TMSCF$_3$ was first prepared in the early eighties by Ruppert,[23] but has only received considerable attention since the discovery of its nucleophilic reactivity toward carbonyl compounds in 1989 by Prakash.[24] Since then, considerable effort has been devoted to the development of different catalytic systems for the activation of the Ruppert-Prakash reagents and include nucleophilic initiators such as fluoride anion (CsF, TBAF, TBAT), alkoxide (KOtBu), amine N-oxide (Me$_3$NO), acetate (LiOAc), N-heterocyclic carbenes (NHC), phosphine (P(tBu)$_3$), as well as electrophilic initiators such as Lewis acids (TiF$_4$/DMF, Cu(OAc)$_2$/dppe/toluene).[14] TMSCF$_3$ is probably the best known and studied of all trifluoromethylating reagents and allows the direct preparation of trifluoromethylated alcohols from ketones and aldhydes, ketones from esters, acetamides from ketones and amines from imines. All these strategies have been extensively reviewed.[14, 25]

Originally, Me$_3$SiCF$_3$ was prepared from CF$_3$Br and hexaethylphosphours triamide in the presence of Me$_3$SiCl in 95% yield. This preparation has a major drawback as it requires the use of ozone depleting and currently prohibited CF$_3$Br. Two alternative methods for the preparation of TMSCF$_3$ without the use of CF$_3$Br have been proposed by Pawelke[26] and Prakash[27]. Based on the work of Pawelke, Dolbier developed an alternative trifluoromethylation method that involves the initial formation of a charge-transfer complex between CF$_3$I and TDAE. Non-enolizable aldehydes, ketones, and aromatic aldimines are trifluoromethylated by this method in moderate to high yields as shown in Scheme 3.[28]

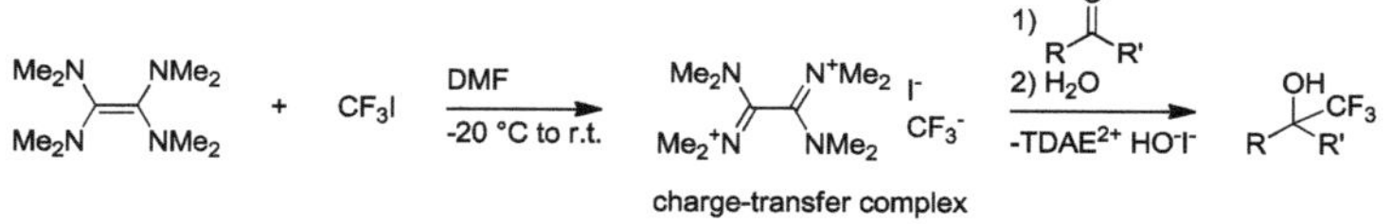

Scheme 3. CF$_3$I/TDAE as alternative CF$_3^-$ source.[28]

Prakash has used phenyltrifluoromethylsulfoxide and sulfone for the reaction with TMSCl to form TMSCF$_3$ as an alternative to the original synthesis of the Ruppert-Prakash reagent.[27] Phenyl trifluoromethyl sulfoxide and sulfone themselves appear to be potential trifluoromethylating reagents upon intiation by nucleophilic activators and efficient trifluoromethylation of nonenolizable carbonyl compounds is observed upon using KOtBu as an initiator (Scheme 4).[29]

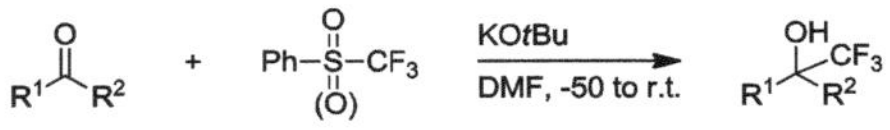

Scheme 4. Phenylsulfoxide and -sulfone as CF$_3$-source for non-enolizable aldehydes and ketones.[29]

Phenyltrifluormethylsulfoxide and sulfone were prepared from diphenyldisulfide and CF$_3$H in the presence of a strong base in DMF followed by the oxidation of phenyl trifluoromethylsulfide intermediated with hydrogen peroxide or *m*CPBA. Trifluoromethane is a cheap and environmentally friendly reagent. Strong bases such as KOtBu are able to deprotonate fluoroform to generate the trifluoromethyl anion and in DMF the anion is stabilized by the formation of an adduct. Based on this concept, Langlois developed stable hemiaminals of fluoral and their silylated derivatives as powerful trifluoromethylating agents towards non-enolizable carbonyl compounds, disulfides and diselenides under activation by a stoichiometric amount of a strong base or catalytic amounts of fluoride anions such as CsF or TBAT as shown in Scheme 5.[30]

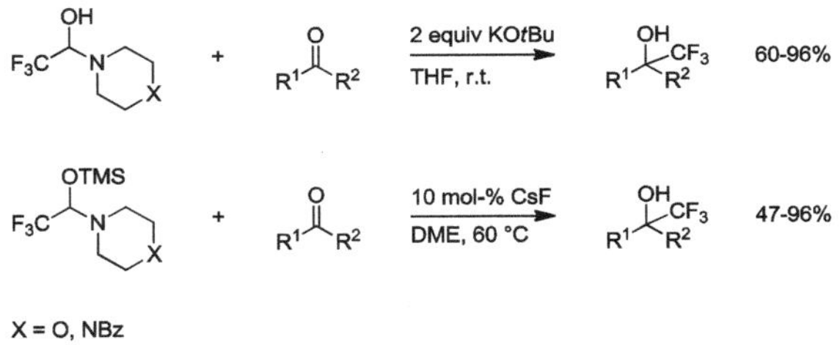

X = O, NBz

Scheme 5. Nucleophilic trifluoromethylation utilizing Langlois' reagents.[30]

However, the use of a strong base precludes the reaction of enolizable substrates. To extend the substrate scope novel trifluoroacetamides and trifluoromethansulfinamides derived from *O*-silylated *vic*-aminoalcohols were prepared and are able to trifluoromethylate both enolizable and non-enolizable ketones, as well as reactive aldehydes in good to excellent yields under fluoride activation at room temperature.[31]

1.3.2 Radical Trifluoromethylation

Dolbier has written an extensive review on the structure, reactivity and chemistry of fluoroalkyl radicals,[32] and the previously cited articles by McClinton[5] and Ma[14] focus on their synthetic applications. Radical trifluoromethylation is the oldest method for the direct transfer of a trifluoromethyl group. Therefore, several radical precursors have been developed and the trifluoromethyl radical can be generated under oxidative, reductive, photochemical, thermal, and electrochemical conditions. Although the trifluoromethylation reaction under radical conditions has been studied extensively due to the availability of precursors and the stability of trifluoromethyl radical species, only two examples have addressed an enantioselective radical trifluoromethylation reaction to date.[33] Mikami and co-workers employed radical trifluoromethylation of lithium enolates, in the presence of (S,S)-hydrobenzoin dimethyl ether or (-)-sparteine to obtain the desired trifluoromethylated products in low yields and ee up to 44%.[34] Despite the relatively low yields and enantioselectivities the results demonstrate the possibility of catalytic asymmetric radical trifluoromethylation of enolates. In 2009, MacMillan reported the first enantioselecitive, organocatalytic trifluoromethylation of aldehydes as shown in Scheme 6.[35] The asymmetric reaction was accomplished via the combination of an imidazolidone organocatalytic cycle and a photoredox catalytic cycle in which the CF_3 radical was generated from the reduction of CF_3I by an iridium photocatalyst under fluoresent light.

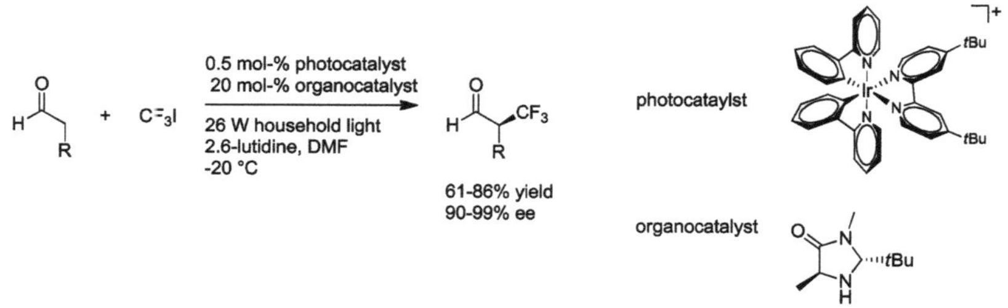

Scheme 6. Radical enantioselective organocatalytic trifluoromethylation of aldehydes.[35]

1.3.3 Electrophilic Trifluoromethylation

Of the three fundamental methods for trifluoromethylation of organic molecules, the electrophilic introduction was, until recently, the least developed. In recent years, however, several new reagents as well as simpler preparations of older reagents have been reported. Due to the high stability and reactivity of these reagents, several are now commercially available. In the last five years, a renaissance of this chemistry occurred

Introduction

as demonstrated in two very recent reviews of this topic by Shibata and Cahard,[3] and by Magnier.[36] Furthermore, an extensive overview on this topic is found in the PhD thesis of R. Koller.[37] Several effective reagents have been developed by the groups of Yagupolskii, Umemoto, Shreeve, Adachi, Magnier, Togni and Shibata and can be divided into three main categories: O-, S-, Se-, Te-(trifluoromethyl) chalcogenium salts I, sulfoxime II and hypervalent iodine(III) reagents III as shown in Figure 1.

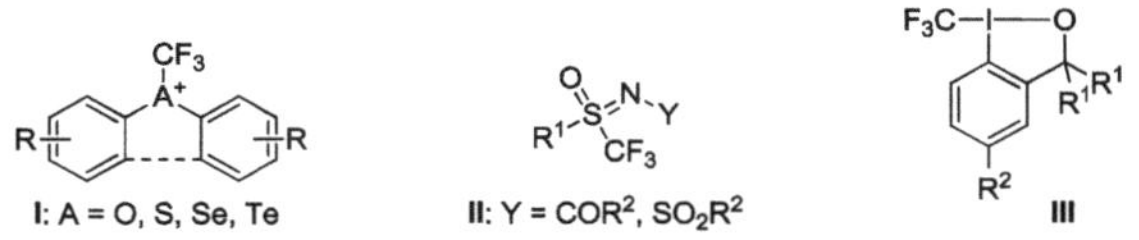

I: A = O, S, Se, Te   II: Y = COR^2, SO$_2$R^2 III

Figure 1. Main categories of electrophilic trifluoromethylating reagents: chalcogenium salts I (left), sulfoxime derivatives II (middle) and hypervalent iodine(III) derivatives III (right).

The first successful electrophilic trifluoromethylation was reported relatively recently, compared to nucleophilic and radical trifluoromethylation, in 1984 by Yagupolskii.[38] This was demonstrated by the effective trifluoromethylation of sodium 4-nitrobenzenethiolate by S-trifluoromethyl diarylsulfonium salt I.

Despite the potential of the newly developed agent, the group did not pursue further studies of the reactivity of this new reagent until 2008.[39] Meanwhile the research groups of Umemoto,[40] Shreeve,[41] Magnier in collaboration with GlaxoSmithKline[42] and Yagupolskii[39] reported improved routes for the preparation of Ar$_2$S$^+$CF$_3$ salts I. These new synthetic approaches allow an easier preparation of the reagents from readily available and cheap starting materials. Furthermore, several new derivatives have been prepared and it was shown that the reactivity is enhanced by the presence of electron-withdrawing groups on the aromatic rings. Not only thiophenolates, but also other soft nucleophiles, i. e. sulfur in sulfinic acid or thiourea, phosphorus in sodium diethylphosphite, iodine in NaI, and carbon in electron rich heterocycles namely, N-methylpyrrole and indole, have been shown to be trifluoromethylated in high yields.[39]

In the early 1990's Umemoto and co-workers achieved a major breakthrough in the field of electrophilic trifluoromethylation by the preparation of heterocyclic analogues of the above discussed diarylsulfides i.e. S-, Se- and Te-(trifluoromethyl)dibenzothio-, -seleno- and -tellurophenium salts I.[40, 43] The relative trifluoromethylating power of these chalcogenium salts increased in the order Te < Se < S while reagents bearing electron-withdrawing substituents showed higher reactivity than derivatives with electron-rich aryl systems.[44] The broad reactivity range of these reagents allowed the trifluoromethylation of a wide range of nucleophiles, including lithium phenylacetylide, activated aromatics, heteroaromatics, enol silyl ethers, enamines, phosphines, thiolates, and sodium iodide.

Introduction

These reagents quickly became the first choice for many chemists as they are not only relatively simple to prepare but also commercially available. Some recent results utilizing S-(trifluoromethyl)dibenzothiophenium salts in the metal catalyzed electrophilic trifluoromethylation include the Pd(II)-catalyzed *ortho*-trifluoromethylation of arenes by C-H activation,[45] the copper-catalyzed trifluoromethylation of aryl- and heteroarylboronic acids[46] and the copper-cataylzed trifluoromethylation of terminal alkenes *via* allylic C-H bond activation.[47] Although Umemoto reported an enantioselective electrophilic trifluoromethylation of a ketone enolate mediated by an chiral borepin in up to promising 42% enantiomeric excess very early on,[48] no progress in this field was reported over the following decade. To date, only two further reports concerning enantioselective electrophilic trifuoromethylation utilizing S-(trilfuoromethyl)dibenzothiophenium salts can be found in the literature: an enantioselective trifluormethylation of methyl 1-oxoindan-2-carboxylate and cinchona ammonium salts acting as chiral phase-transfer catalyst,[14] and an enaticselective trifluoromethylation of β-keto esters induced by as chiral guanidine in up to 71% ee.[49]

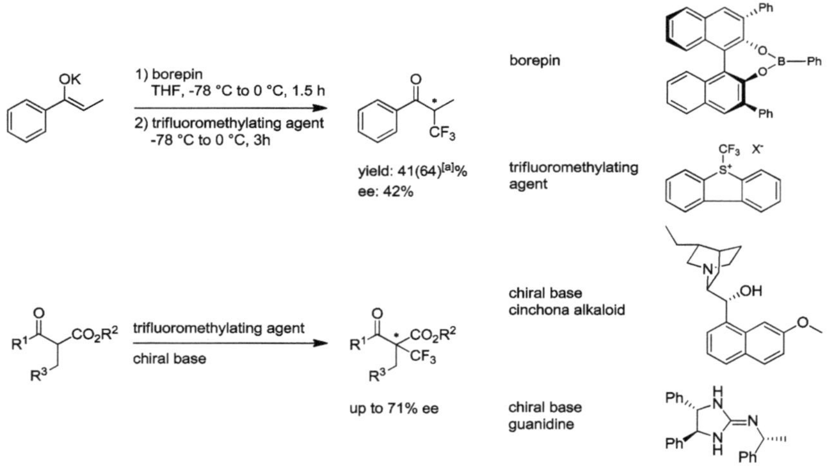

Scheme 7. Enantioselective electrophilic trifluoromethylation, utilizing S-(trifluoromethyl)dibenzothiophenium salts, of ketone enolate (upper),[48] β-keto esters (lower).[14, 49] [a] Yield determined on the basis of ^{19}F NMR integration using $C_6H_5CF_3$ as internal standard.

In 2007, Umemoto published the *in situ* preparation of unstable O-(trifluoromethyl)-dibenzo furanium salts, completing the series of non-radioactive chalcogens with the most electro negative one, oxygen, giving the most reactive reagent.[50] Following a photoirradiation protocol at −100 °C allows the *in situ* preparation of the trifluoromethylating agent, which is able to trifluoromethylate amines, anilines, and pyridines under very mild conditions.

Recently, a new class of S-(trifluoromethyl)sulfonium salts I was developed by Shibata and co-workers.[51] S-(trifluoromethyl)-2-cyclopropylthiophenium triflates showed enhanced reactivity towards β-ketoesters and dicyanoalkylides compared to other derivatives, S-(trifluoromethyl)dibenzothiophenium salts and hypervalent iodine(III) reagents (*vide infra*).

A second class of electrophilic trifluoromethylating agents is based on S-trifluoromethyl sulfoximes II. Cyclic as well as acyclic derivatives have been reported in a Japanese patent by Adachi and Ishihara.[52] Soft nucleophiles such as thiolates and enamines are trifluoromethylated in moderate yields. In addition, theses reagents show remarkable reactivity towards hard nucleophiles such as Grignard reagents and lithium acetylides. In 2010, Shibata and co-workers, published the synthesis and application of [(oxido)phenyl(trifluoromethyl)-λ^4-sulfanylidene]dimethylammonium tetrafluoroborate,[51] the trifluoromethyl derivative of Johnson's methylating agent.[53] In the presence of a base, β-keto esters, as well as dicyanoalkylidenes were trifluoromethylated in good to high yields.

The previously described reagents are all based on a trifluoromethyl group attached to a polyvalent chalcogen. A completely different approach was chosen in our research group. Research was focused on trifluoromethylating reagents III based on hypervalent iodine(III) utilizing the highly electron deficient, Lewis acidic character of such compounds.

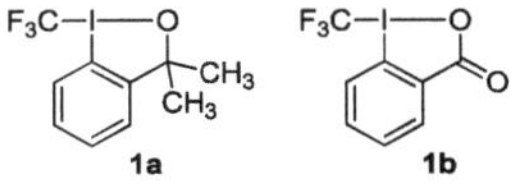

Figure 2. The two mainly used trifluoromethylating agents based on hypervalent iodine.

An introduction to hypervalent iodine compounds, as well as the synthesis of these reagents is given in Section 2.1. Prior to the work described in this dissertation, reagent **1a** and **1b** have shown to react smoothly with a variety of *C*-,[54] *S*-,[54a, 55] *P*-,[56] and *O*-[57] centered nucleophiles as shown in Figure 3.

Furthermore, these reagents have been utilized by several other research groups especially, in the rapidly growing field of transition-metal-catalyzed electrophilic trifluoromethylation. Some recent results include the copper-catalyzed trifluoromethylation of unacitvated terminal olefins,[58] alkynes,[59] (hetero)aryl- and alkenylbronic acids,[60] and indols.[61] Potassium vinyltrifluoroborates were successfully trifluoromethylated under iron(II) catalysis.[62] Trifluoromethylated arenes can be directly produced by *in situ*

Introduction

iridium-catalyzed C-H activated borylation followed by copper-catalyzed trifluoromethylation.[63] Sanford and co-workers used these reagents for the preparation of a monomeric F_3C-Pd(IV) aquo complex. Furthermore, MacMillan reported the highly enantioselective, organocatalytic α-trifluoromethylation of aldehydes by the hypervalent iodine compound **1a**. To prevent racemization, the aldehydes formed were reduced *in situ* to the corresponding trifluoromethylated alcohols.[64] A diastereoselective approach to chiral α-trifluoromethylated alcohols and acids has been presented by our group in collaboration with Cahard.[54d] The configuration at the tertiary carbon atom formed during the trifluoromethylation of the carbonyl system was controlled with Evans-type oxazolidones. The diastereoisomers were formed in up to 91% combined yield and 97:3 dr. The isolated diastereopure products could be transformed without racemization to the corresponding alcohols and acids.

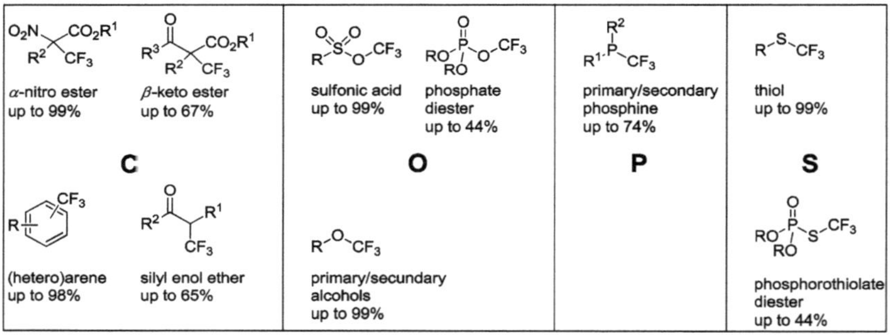

C		O		P	S
α-nitro ester up to 99%	β-keto ester up to 67%	sulfonic acid up to 99%	phosphate diester up to 44%	primary/secondary phosphine up to 74%	thiol up to 99%
(hetero)arene up to 98%	silyl enol ether up to 65%	primary/secundary alcohols up to 99%			phosphorothiolate diester up to 44%

Figure 3. Products of electrophilic trifluoromethylation of various nucleophiles by hypervalent iodine(III) reagents.[54-57]

As summarized in this chapter, there has been a massive development in the trifluoromethylation of organic molecules over the past several years. Some especially remarkable progress has been achieved in the field of electrophilic trifluoromethylation. Despite these improvements, there are still some challenges; especially the direct electrophilic trifluoromethylation of hard nucleophiles, such as amines, phenols and alcohols, remain targets for the future. Chapter 2 will address these problems *via* a correlation of the structures and reactivities of hypervalent iodine(III) reagents, whereas in Chapter 3 the synthesis of *N*-trifluoromethylated compounds will be described.

2 Structure and Reactivity Correlation of Hypervalent Iodine Reagents for Electrophilic Trifluoromethylation

2.1 Introduction

Since the early days of modern chemistry, hypervalent compounds such as ICl_3, prepared by Gay-Lussac in 1814,[65] have been synthesized. However, the term "hypervalent" was not coined until 1969, when Musher recommended its use to describe molecules in which atoms exceed the number of valence electrons allowed by the traditional octet rule formulated by Lewis and Langmuir.[66] The most common hypervalent iodine compounds are aryl-λ^3-iodanes (ArIL$_2$) with pseudotrigonal bipyramidal, also called T-shaped geometries and aryl-λ^5-iodanes (ArIL$_4$) with square pyramidal coordination geometries. Since the discovery of the first organic hypervalent iodine compound, (dichloro-λ^3-iodo)benzene, by Willgerodt,[67] various derivatives have been prepared and found applications in diverse fields. Possibly the most renowned member of this class of compounds is the Dess-Martin periodane (DMP), a mild selective oxidant for the conversion of primary and secondary alcohols to the corresponding aldehydes and ketones, respectively.[68] Furthermore these hypervalent organoiodine compounds find widespread application in organic synthesis, including fragmentation and rearrangement reactions and they are also used as formal electrophilic group transfer reagents.[69] For instance, the more stable and efficient derivatives of the first generation of perfluoroalkylating agents based on hypervalent iodine(III) developed by Yagupolskii,[70] (perfluoroalkyl)aryliodonium triflates (FITS) and hydrogensulfonates (FIS), are used with a variety of nucleophiles, from simple inorganic salts to organic substrates.[44] Despite their success, derivatives with the simplest perfluoroalkyl unit were not accessible, and it has been speculated that the desired compounds are unstable. The formation of CF_3I (δ_F = –7.5 ppm) observed by ^{19}F NMR spectroscopy by the reaction of (difluoroiodo)toluene and TMSCF$_3$ under fluoride catalysis affirmed this assumption. These results suggested that for additional stability, a more rigid, and therefore stable, backbone is necessary. Heterocyclic iodanes have a considerably higher stability than their acyclic analogs, this effect is normally explained by the bridging of an apical and an equatorial ligand and a better overlap of the lone pair electrons on the iodine atom with the π-orbitals of the benzene ring.[71]

In 2006, our group succeeded in preparing of several hypervalent organoiodine(III) compounds bearing a trifluoromethyl group and based on a benziodoxole scaffold.[7, 54b] Rather than utilizing aromatic substitution chemistry, as in the previous attempts by Yagupolskii and Umemoto, the I-CF$_3$ bond was constructed in an umpolung sequence between a suitable trifluoromethyl anion source and an I(III) fragment. The synthesis of these trifluoromethylating agents **1a-d** is shown in Scheme 8 and involves three fundamental steps: oxidation to the cyclic iodine **2a-d**, ligand exchange to a more reactive intermediate **3a-d** and in the last step an umpolung of the nucleophilic CF_3-source (Ruppert-Prakash reagent) in the presence of fluoride. As pointed out in Section 1.3.3 reagents **1a-d** can be applied to a variety of nucleophiles, whereby during the reaction the reagent is reduced to the corresponding iodo-alcohol **4a-d**, a formally recyclable CF_3-carrier.

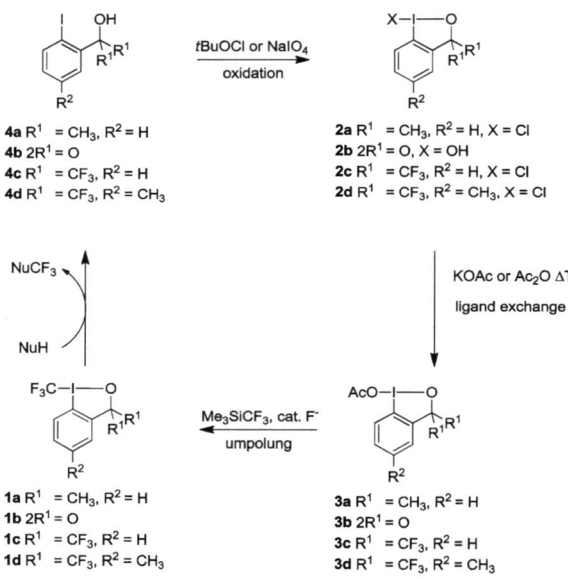

Scheme 8. Synthesis of electrophilic trifluromethylating agents based on hypervalent iodine.

At the time this project was initiated, positive results for the electrophilic trifluoromethylation of a multitude of substrates, especially soft nucleophiles, had already been obtained. Even so, the reagents had failed to trifluoromethylate hard nucleophiles such as oxygen- and nitrogen-centered ones. It was speculated that the reactivity and selectivity of the reagents may be fine-tuned by altering the iodoxole backbone; steric as well as electronic factors can be varied by functionalization of the aromatic core, altering substituents on the heterocyclic ring, changing supporting ligand from oxygen to another heteroatom and going from a neutral to a cationic species.

2.2 Synthesis of Derivatives

In the very first report of the synthesis of hypervalent iodine compounds for electrophilic trifluoromethylation, several derivatives of the reagents were reported.[54b] Since then, a variety of new precursors as well as some new reagents have been prepared and their synthesis is discussed in this chapter. In general, they were prepared by the same fundamental approach described in Scheme 8 and therefore 1-chloro-λ^3-iodanes such as **2a** and **2c** are useful intermediates.

2.2.1 Synthesis of 1-Chloro-λ^3-iodanes

Alcohols **5-10** were converted to the neutral 1-chloro-λ^3-iodanes **11-16** by a slight excess of tBuOCl as oxidant in DCM and the results of these oxidations are shown in Table 1.

Table 1. Synthesis of neutral, five-memberd heterocylic 1-chloro-λ^3-iodanes.

5	66%	11			95%
6	82%	12			92%
7	24%[a]	13			58%
8	80%[b]	14			82%[c]
9	not det.[d]	15			44%[d]
10[e]		16			89%

[a] Prepared without the addition of CeCl$_3$; [b] prepared according to reference;[72] [c] prepared according to reference in CCl$_4$;[73] [d] **9** was not isolated in pure form and directly converted to **15** in 44% overall yield over two steps; [e] synthesized according to Scheme 9.

Structure and Reactivity

The alcohols needed for the oxidation step are derived from the corresponding ketone and the addition of a Grignard reagent. Alcohol **10** was synthesized in 7 steps from dicyclopentadiene (**17**) according to Scheme 9. In the first three steps, dicyclopentadiene was converted to ethyl cyclopentadiene carboxylate. In the following two steps, ester **18** was synthesized according to the reported method of Tanida and Irie.[74] The tertiary alcohol **19** was prepared by a twofold Grignard addition to the ester under reflux in high yield and converted to **20** by *ortho* lithiation with sBuLi/TMEDA followed by addition of 1,2-diiodoethane.

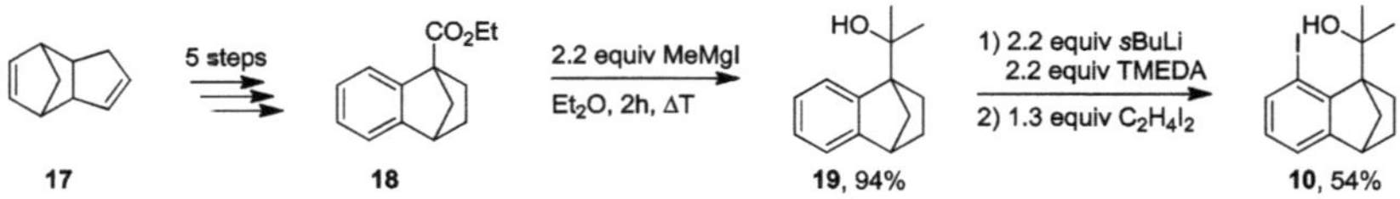

Scheme 9. Synthesis of **10** in seven steps from dicyclobutadiene.

As pointed out in the end of this chapter's introduction one of the possibilities to tune reagent activity is changing the supporting ligand from oxygen to another heteroatom. Furthermore, a cationic species instead of a neutral compound should show enhanced electrophilicity. According to Scheme 9, the cationic 1-chloro-λ^3-iodanes **20** and **21** were prepared from oxazoline **22** and **23** by oxidation with tBuOCl after protonation with HBF$_4$. The protonation before the oxidation is essential and under these slightly modified conditions **20** and **21** are obtained in high yields.

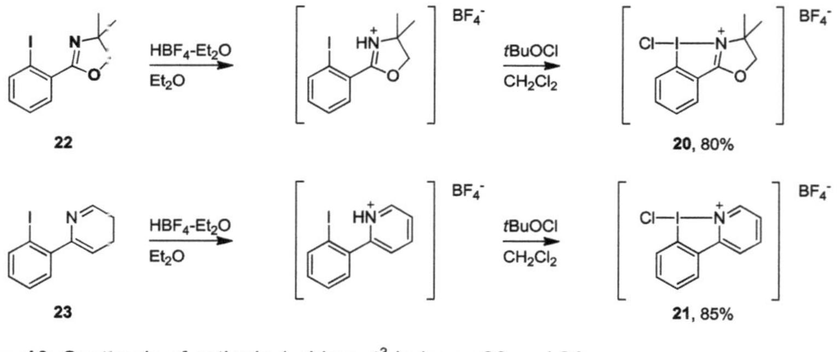

Scheme 10. Synthesis of cationic 1-chloro-λ^3-iodanes **20** and **21**.

Another equally interesting derivative of 1-chloro-λ^3-iodanes would be a hypervalent iodine species stabilized by a fluorine atom analogous to the above shown oxygen and nitrogen stabilized hypervalent iodine(III) compounds. Such an intramolecular fluorine stabilization would be an unprecedented interaction.

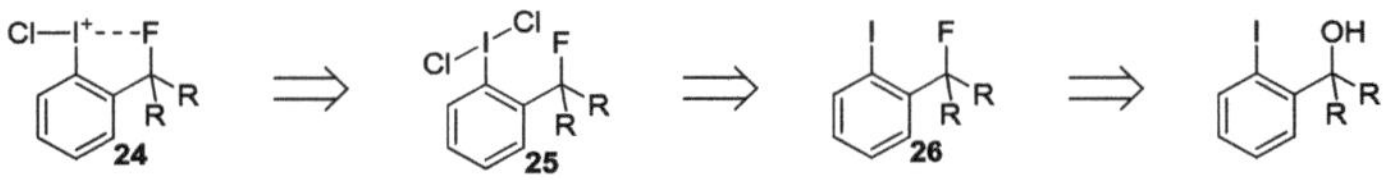

Structure and Reactivity

Scheme 11. Retrosynthesis of target 1-chloro-λ^3-iodanes.

The synthesis of 1-chloro-λ^3-iodanes **24**, where the iodine(III) center is stabilized by an I-F interaction, was proposed following similar considerations as in the previous cases. The retrosynthetic analysis is shown in Scheme 11. The target molecule should be accessible by chloride abstraction from a 1-dichloro-λ^3-iodane **25**, which is derived by oxidation from the corresponding fluorinated compound **26**. The fluorine might be introduced by deoxofluorination. Scheme 12 shows the successful synthesis of 1-dichloro-λ^3-iodanes **25a** and **25b**.

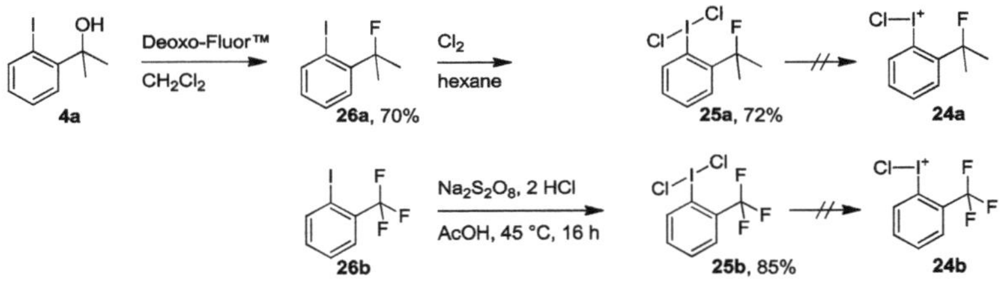

Scheme 12. Synthetic attempts towards 1-chloro-λ^3-iodanes **24a/b** stabilized by an adjacent fluorine atom.

Compound **4a** was fluorinated with Deoxo-Fluor™ analogously to the synthesis described by Cheng,[9b] whereby water elimination occurred as side reaction. Depending on the substrate, iodoarenes were either oxidized with sodium peroxodisulfate and HCl, or with Cl_2 in an apolar solvent as reported by Klimaszewska[75] and Gladysz,[76] respectively. The hypervalent compound **25a** can be recrystallized via slow diffusion of pentane into a saturated CH_2Cl_2 solution and single crystals suitable for X-ray structure determination were obtained. The ORTEP representation of **25a** is shown in Figure 4. The 1-fluoro-1-methylethyl-group has a slightly twisted staggered conformation (torsion angles: C^1-C^6-C^7-F^1 -41.1(5)° and C^1-C^6-C^7-C^8 75.7(5)°). The dichloro-λ^3-group is tilted slightly away from the fluorinated ortho substituent, as indicated by the C^6-C^1-I^1 angle of 124.3(3)°. These findings indicate that there is no additional stabilization by the fluorine atom, despite the short F-I distance of 3.021(2) Å.

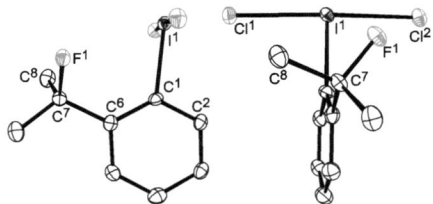

Figure 4. ORTEP drawing of X-ray structure of 1-(dichloro-λ^3-iodanyl)-2-(1-fluoro-1-methylethyl)benzene (**25a**). Hydrogen atoms are omitted for clarity, thermal ellipsoids set to 50% probability. Selected bond lengths [Å], bond angles [°] and torsion angles [°]: I^1-Cl^1 2.531(1), I^1-Cl^2 2.456(1), C^1-I^1 2.117(4), I^1-C^1-C^6 124.4(3), Cl^1-I^1-Cl^2 177.63(3), C^1-C^6-C^7-C^8 75.7(5), C^1-C^6-C^7-F^1 -41.1(5).

Chloride abstraction was attempted using halide scavengers such as $KSbF_6$, $SbCl_5$, $AgBF_4$, $TIPF_6$, $Et_3O^+BF_4^-$ and TMSOTf in various solvents, but in all cases the only products observed by NMR were fully reduced species and Cl_2I^+ salts.

2.2.2 Synthesis of Hypervalent Trifluoromethyl Compounds

Following the original reaction sequence for the preparation of 1-trifluoromethyl-λ^3-iodanes as outlined in Scheme 8 compounds **27**, **28** and **29** were obtained in two steps from their 1-chloro-λ^3 derivatives and the yields are given in Table 2.

Table 2. Synthesis of 1-trifluoromethyl-λ^3-iodanes **27-29**.

Entry	Product		Acetate Source	Yield [%]
1	**27**	F_3C-I-O (cyclohexyl)	AgOAc	37
2	**28**	F_3C-I-O (Ph)	KOAc	81
3	**29**	F_3C-I-O	KOAc	14

All syntheses were carried out without isolating the acetate intermediate in CH_3CN with an excess of $TMSCF_3$ and TBAT as catalytic fluoride source.

Under these conditions the cationic 1-chloro-λ^3-iodane **20** and **21** could not be transformed to the desired products, since difficulties were encountered in the ligand exchange step, and inseparable product mixtures were obtained.

Excursion

Oxygen trifluoromethylated compounds are important building blocks for the pharmaceutical industry, but are difficult to access synthetically. With the goal of developing a convenient methodology for electrophilic trifluoromethoxylation, the synthesis of a series of hypervalent iodine(III) based reagents bearing an OCF_3 group was suggested. Reagents of this type might be constructed along similar lines to those described for the trifluoromethylating reagents discussed in this chapter, i. e. by the reaction of a suitable hypervalent iodine(III) precursor and a nucleophilic OCF_3 source. Nucleophilic trifluoromethanolate salts are historically hard to access. In 2008, Kolomeitsev[77] as well as a Merck patent[78] described their synthesis from trifluoromethyltriflate (TFMT) and anhydrous F^- anion sources, instead of CF_2O[79] or CF_3OCl[80] which were utilized in earlier reports. When the 1-chloro-λ^3-iodanes **2a** or **2c** were allowed to react with $AgOCF_3$, prepared *in situ* from AgF and TFMT, the corresponding 1-fluoro-λ^3-iodanes **30a** and **30b**, respectively, were formed instead.

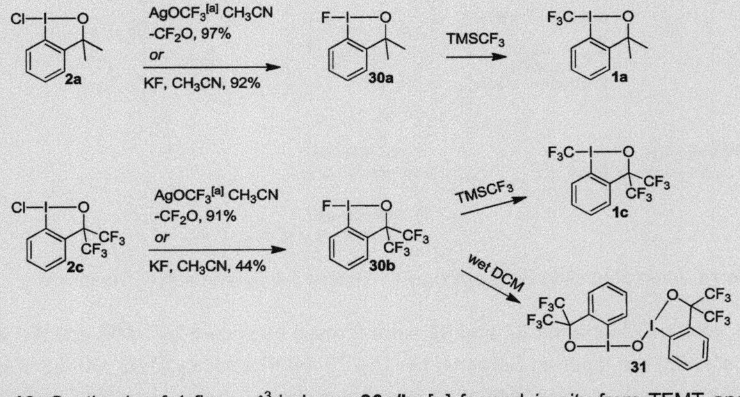

Scheme 13. Synthesis of 1-fluoro-λ^3-iodanes **30a/b**; [a] formed *in situ* from TFMT and AgF in CH_3CN at -30 °C.

Fluorophosgene was observed by ^{19}F NMR spectroscopy (δ_F = –19.3 ppm) as a byproduct of these experiments, suggesting that the desired product was most likely formed, but decomposed under CF_2O elimination to the 1-fluoro-λ^3-iodanes **30a** and **30b**, respectively. The newly formed 1-fluoro-λ^3-iodanes **30a/b** are sensitive to moisture and hydrolyze over time. When **30b** was recrystallized from wet dichloromethane the μ-oxo-bridged **31** was formed as confirmed by X-ray analysis and HRMS. Compounds **30a/b** can also be obtained by ligand exchange from their 1-chloro derivatives **2a/c** upon addition of KF in acetonitrile and react with $TMSCF_3$ slowly to yield the trifluoromethylating reagents **1a** and **1c**.

Structure and Reactivity

To circumvent the above described ligand-exchange problems, the synthesis of cationic hypervalent iodine(III) compounds was attempted. In order to obtain the desired compounds, the synthetic approach was modified such that **22** and **23** were directly oxidized to the corresponding 1-fluoro iodanes **32** and **33** which can be trifluoromethylated directly without exchanging the ligand as outlined in Scheme 14.

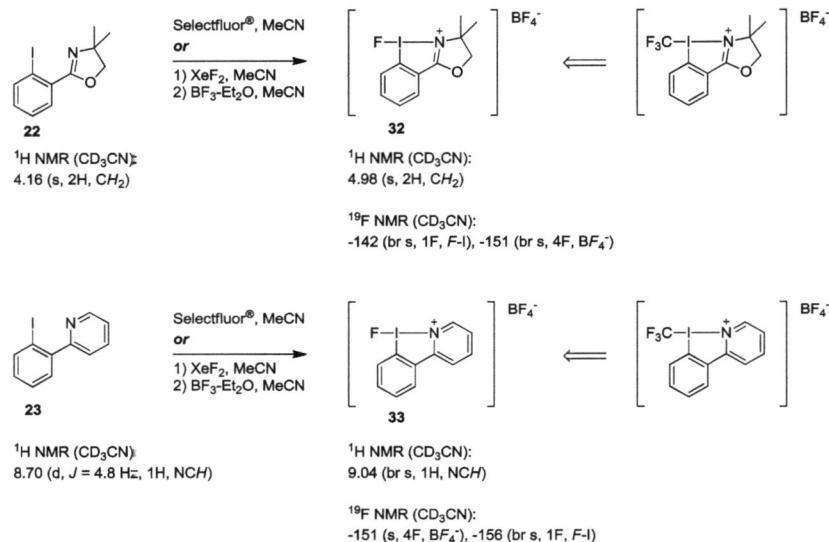

Scheme 14. Attempted alternative pathway to cationic 1-trifluoromethyl-λ^3-iodanes.

The 1-fluoro-λ^3-iodanes **32** and **33** were formed as shown by NMR and the important chemical shifts are given in Scheme 14. The ^{19}F NMR spectra of **32** and **33** show broad linewidths for both signals indicating fluorine exchange between iodine bound fluorine and BF_4^- ion. In addition, the signal corresponding to the fluorine bound to the iodine decreases in intensity over time, which might be explained by hydrolysis and formation of HF. Characterization by methods other than NMR was not possible and isolation of **32** and **33** proved equally challenging. When Selectfluor® was used as fluorinating agent, the product could not be separated from the reaction mixture. No reproducible results were obtained for the reaction with XeF_2 and BF_3-Et_2O. Other fluorinating agents or changing the reaction conditions such as temperature or the mode of addition did not lead to clearer results.

Probably the simplest of all trifluoromethylating agents based on hypervalent iodine would be a $(CF_3)_2I^+$ salt, since the methylated analog $(CH_3)_2I^+$ is known to react as methylating agent.[81] Scheme 15 shows the synthetic attempts towards the potential new trifluoromethylating agents.

$$\underset{F}{\overset{O}{\|}}{\overset{}{\underset{F}{\bigvee}}} + ClF \xrightarrow{CsF} CF_3OCl \xrightarrow[-COF_2]{CF_3I} Cl\text{—}I\text{—}F \underset{CF_3}{} \xrightarrow[\text{additives}]{Me_3SiCF_3,\ Me_3SnCF_3,\ or\ Hg(CF_3)_2} F_3C\text{—}\underset{Cl}{I}\text{—}CF_3 \xrightarrow{MX_n} \underset{F_3C\text{—}I^+\text{—}CF_3}{MX_nCl^-}$$

Scheme 15. Attempts of the synthesis of $(CF_3)_2I^+$.

Due to the weakened I-F bond, $CF_3I(Cl)F$ can be transformed into $(CF_3)I(OCH_3)Cl$ upon reaction with trimethylmethoxysilane.[82] This enhanced reactivity, that allows the selective exchange of only one halogen, was planned to be utilized in the synthesis of $(CF)_3I^+$. $CF_3I(Cl)F$ is formed upon oxidation of CF_3I with CF_3OCl and subsequent fluorophosgene elimination.[83] Unfortunately, $CF_3I(Cl)F$ undergoes halogen scrambling in solution to form CF_3IF_2 and CF_3ICl_2 even at low temperatures and did not react with neat Me_3SiCF_3 or Me_3SnCF_3 at temperatures slightly above their melting point. Neither the addition of additives such as TBAT, Me_3NO, CsF, NaI, KOAc with or without various solvents nor the use of $Hg(CF)_3$ as trifluoromethylating agent led to any success. The I-F bond could not be cleaved at low temperature, and at higher temperatures the formation of CF_3I was observed by ^{19}F NMR spectroscopy (δ_F = −7.5 ppm) indicating decomposition of the hypervalent species.

2.3 Solid State Structure Analysis

2.3.1 X-Ray Structures of 1-Chloro-λ^3-iodanes

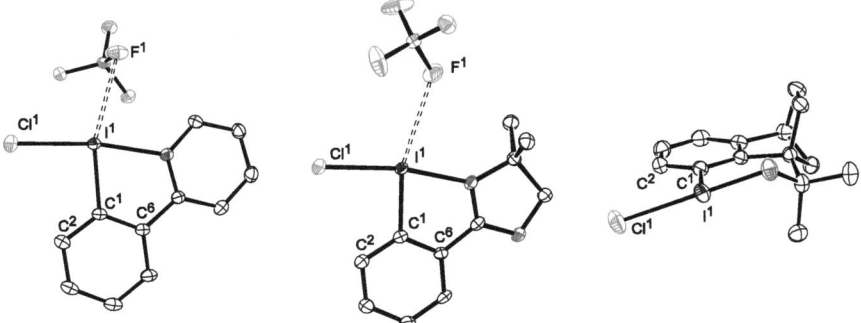

Figure 5. ORTEP drawing of X-ray structures of **21** (left), **20** (middle) and **16** (right). Hydrogen atoms are omitted for clarity, thermal ellipsoids set to 50% probability

Crystals for X-ray analysis of compound **2a** were obtained by slow evaporation of a dichloromethane solution. Compounds **11-16** were crystallized from CH_2Cl_2 by diffusion of pentane or Et_2O into the saturated solution. Under these conditions **14** crystallizes in the space group $P\bar{1}$ incorporating a solvent molecule. However, crystalliztion from EtOAc solutions leads to solvent free crystals in space group $P2_1/c$. Single crystals for X-ray analysis of the cationic compounds **20** and **21** were obtained by diffusion of Et_2O and CH_2Cl_2, respectively, into saturated acetonitrile solutions at −18 °C. Representations of the structures are shown in Figure 5.

Previous to this work, only the four crystal structures of 1-monochloro-λ^3-iodanes **2c** and **34-36** (Figure 6) were known.[84]

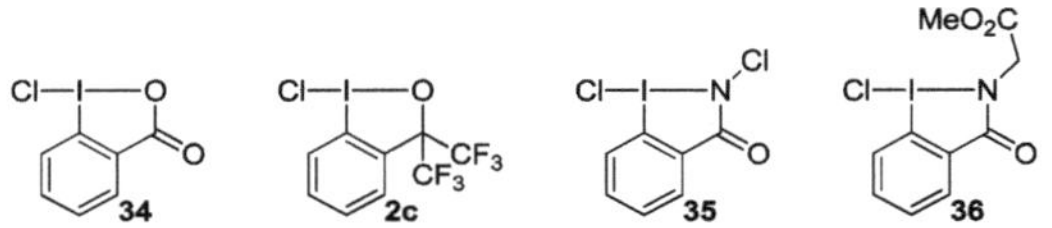

Figure 6. 1-Chloro-λ^3-iodanes with previously published crystal structures.

Table 3 lists important bond lengths, bond and torsion angles of all 1-chloro-λ^3-iodane derivatives synthesized in this work as well as those of **2c, 34-36**.

The crystal structures of all 1-chloro-λ^3-iodanes clearly show a distorted T-shaped geometry around the iodine, typical for members of the hypervalent iodine(III) class. In all five-membered heterocycles the Cl-I-O and the Cl-I-N angles are significantly smaller than 180°. According to the VSEPR model this is due to the repulsion of the two lone pairs at the iodine. The six-membered heterocyclic ring of **16** has a half-chair-type conformation (Figure 5), with the two methyl groups in axial and equatorial positions and shows an almost perfect T-shaped geometry around the iodine atom, as indicated by the Cl-I-O angle of 178.87(4)°. In compound **34**, the oxygen in the five-membered heterocycle is almost perfectly coplanar to the adjacent π system (O-I-C^1-C^6 torsion angle 1.8°). Through the series **34-2c-13-11-2a-12-14-15-12** the O-I-C^1-C^6 torsion angles rise indicating that the five-membered heterocyclic ring displays an envelope-type conformation with the oxygen out of the π plane, as illustrated Figure 7.

Table 3. Important bond lengths, bond and torsion angles of **2a/c, 11-16, 20/21, 34-36**.

Bond lengths [Å]	Cl-I	O/N-I	C^1-I
2a	2.5491(8)	2.042(2)	2.102(3)
2c[a]	2.438(2)	2.110(5)	2.105(7)
34[a]	2.461(1)	2.091(3)	2.100(4)
11	2.5135(9)	2.049(2)	2.108(3)
12[b]	2.5751(13) /2.5741(13)	2.016(3) / 2.005(3)	2.104(5) / 2.113(5)
13	2.5201(7)	2.0511(18	2.117(2)
14[c]	2.5406(7)	2.0437(2)	2.107(2)
15	2.5805(6)	2.0169(14)	2.108(2)
16	2.5703(6)	2.0220(15)	2.1259(19)
35[d]	2.56	2.06	2.19
36[e]	2.563	2.113	2.101
20	2.4612(9)	2.190(3)	2.112(3)
21	2.4406(4	2.2273(12)	2.1054(14)
Bond angles [°]	N/O-I-Cl	N/O-I-C^1	C^1-I-Cl
2a	171.06(7)	80.57(10)	91.45(8)
2c[a]	172.0(1)	78.9(2)	93.2(2)
34[a]	171.96(8)	79.5(1)	92.6(1)
11	170.17(7)	80.67(10)	90.77(9)
12[b]	170.10(9) / 170.94(10)	79.68(15) / 79.05(16)	90.68(13)/ 92.23(14)
13	170.51(5)	80.20(9)	90.43(7)
14[c]	169.86(5)	79.89(9)	91.15(7)
15	171.61(5)	80.5(7)	91.27(6)
16	178.87(4)	88.86(7)	92.02(5)
35[d]	171	80	90
36[e]	170.6	79.0	91.6
20	169.15(7)	77.56(11)	91.65(9)
21	171.12(3)	77.36(5)	94.21(4)
Torsion angles [°]	N/O-I-C^1-C^6	Cl-I-C^1-C^2	
2a	10.8(2)	6.0(3)	
2c[a]	−6.4(6)	−4.7(5)	
34[a]	1.8(3)	3.3(3)	
11	10.0(2)	2.9(3)	
12[b]	−12.4(3) / 18.0(3)	−11.4(4) / 12.4(4)	
13	8.36(18)	3.9(2)	
14[c]	15.82(17)	13.1(2)	
15	17.17(15)	15.18(18)	
16	13.60(14)	12.74(15)	
35[d]	na	na	
36[e]	1.17	1.96	
20	−1.1(2)	−1.8(3)	
21	−1.01(10)	−4.27(12)	

[a] Data from literature;[85] [b] asymmetric unit contains two independent molecules; [c] data of crystallization in $P2_1/c$; [d] data from literature;[86] [e] data from literature.[87]

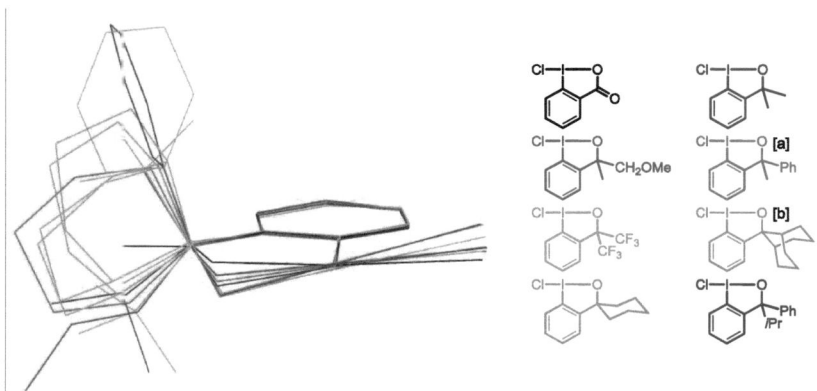

Figure 7. Structural overlap of compounds **2a/c**, **11-15** and **34** generated with the program MERCURY. [a] structure given for the $P2_1/c$ modification; [b] both independent molecules in the asymmetric unit are given.

In the six-membered heterocyclic compound **16** the oxygen atom deviates from the plane to a similar extent as in compound **12** (Cl-I-C^1-C^6 torsion angles of 13.60(14) and −12.4(3)°, respectively). In compounds **20** and **21** the nitrogen sits almost perfectly in the plane of the adjacent system (torsion angle −1.1(2) and 1.01(10)°, respectively), comparable to compound **34**.

The Cl-I bond lengths lie between 2.438(2) and 2.5805(6) Å and are elongated in the order **2c/34-11-13-2a/14-12-15**, while the I-O bond length (2.110(5) to 2.0169(14) Å) decreases, which can mainly be ascribed to electronic effects derived from the substituents. The fine ordering is probably due to packing effects and remote intermolecular contacts in the solid state. The cationic compounds **20** and **21** show no intermolecular contacts, but ion pairing; secondary I-F interactions between the iodine(III) and BF_4^- counterions are observed ranging between 3.028(3) and 3.173(1) Å. The Cl-I bond in the six-membered heterocyclic compound **16** is 2.5703(6) Å, rather long compared to the Cl-I bond lengths in the five-memberd ring compound. The Cl-I bonds in the nitrogen heterocycles are comparable to the short Cl-I-bond in **2c** and **34** (Cl-I bond lengths 2.438(2) and 2.461(1) Å, respectively). In contrast, the I-N bonds are notably longer in comparison to the I-O bonds, also in comparison to the neutral compounds **35** and **36** (2.06 and 2.113 Å, respectively). While the I-heteroatom bond lengths depend on the substituents, the I-C^1 bond lengths remain constant with the exception of compound **13** and compound **16**. The six-membered ring adopts a half-chair conformation and the I-C^1 bond is therefore elongated.

2.3.2 X-Ray Structures of 1-(Trifluoromethyl)-λ^3-iodanes

Single crystals of **28** and **29** were obtained upon cooling of a saturated pentane solution and **27** forms single crystals upon sublimation under reduced pressure at ambient temperature. The corresponding ORTEP representations are shown Figure 8, and Table 4 combines important bond lengths, bond and torsion angles of the trifluoromethylating agents.

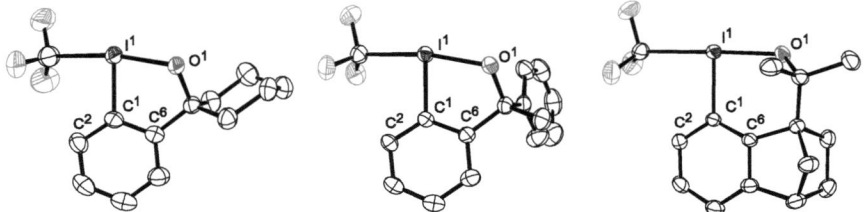

Figure 8. ORTEP drawing of X-ray structures of **27** (left), **28** (middle) and **29** (right). Hydrogen atoms are omitted for clarity, thermal ellipsoids set to 50% probability.

Table 4. Important bond lengths, bond and torsion angles of trifluoromethylating agents.

Bond lengths [Å]	I-CF$_3$	I-O	I-C^1
1a[b]	2.267(2)	2.1176(14)	2.1211(19)
1b[a]	2.219(4)	2.283(2	2.113(3)
1c[a]	2.229(2)	2.2014(15	2.115(2)
1d[a]	2.236(2	2.1977(17	2.114(2)
27	2.262(4)	2.121(2)	2.123(3)
28	2.2580(14)	2.1380(10)	2.1153(12)
29	2.304(2)	2.0979(14	2.1314(18)
Bond angles [°]	O-I-CF$_3$	C^1-I-O	C^1-I-CF$_3$
1a[b]	169.78(7)	78.71(6)	91.11(8)
1b[a]	170.49(12	76.79(11)	93.74(14)
1c[a]	169.40(7)	77.07(7)	92.37(8)
1d[a]	171.07(8)	77.58(7)	93.61(9)
27	169.69(13)	78.15(11)	91.56(15)
28	170.17(5)	78.44(4)	91.75(5)
29	177.90(6)	87.23(6)	91.06(7)

continued on the next page

Structure and Reactivity

Torsion angles [°]	O-I-C¹-C⁶	F₃C-I-C¹-C²
1a[b]	11.56(15)	13.21(18)
1b[a]	0.3(2)	0.7(3)
1c[a]	−12.11(15)	−11.75(19)
1d[a]	3.21(16)	4.3(2)
27	−15.4(2)	−14.7(3)
28	−10.84(8)	−14.45(11)
29	14.94(14)	16.15(14)

[a] Values taken from literature;[7, 54b] [b] values taken from literature.[7]

The same trends observed for the 1-chloro-λ^3-iodanes can be found in the corresponding trifluoromethylating agents, though in a less pronounced manner. Figure 9 shows a bond length correlation of 1-chloro- and 1-(trifluoromethyl)iodanes. The I-O bonds in the trifluoromethylated compounds are longer than the corresponding 1-chloro-λ^3-iodanes and have also shorter iodine bond lengths to the 1-λ^3-substituent. Similarly to the 1-chloro-λ^3-iodanes, a short I-O bond leads to an elongated F₃C-I bond and an almost unchanged C¹-I bond.

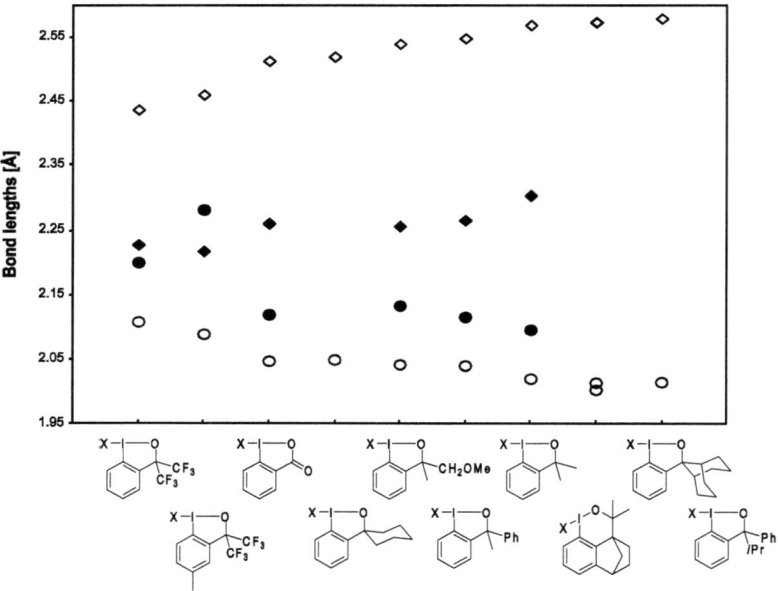

Figure 9. Correlation of bond lengths of 1-chloro- and 1-(trifluoromethyl)iodanes. Unfilled: 1-chloro-λ^3-iodanes, filled: trifluoromethylating agents, squares: O-I bonds, dimonds: I-Cl and I-CF₃, respectively

Structure and Reactivity

The majority of the trifluoromethyl derivatives pack in a similar manner to their 1-chloro-λ^3-precursors and therefore show similar intermolecular contacts. If analogous packing is not observed, bond length comparisons should be made with care. However, our results show that the CF_3 derivatives of corresponding 1-chloro-λ^3-iodanes with long Cl-I bonds are likely to have long F_3C-I bonds as well.

2.4 Reactivity Study

With the wealth of structural information available, studies were undertaken to correlate the structural features of the 1-trifluoro-λ^3-iodanes to their relative reactivities toward a standard substrate to identify structural features, which are potentially advantageous with regard to the CF_3-group transfer and allow the development of improved reagents. As shown in Scheme 16 the initial rates of reaction (v_0) were measured for mixtures of 1-trifluoro-λ^3-iodanes **1a-c** and **27-29** (0.1 M) and toluenesulfonic acid monohydrate (0.1 M) in a 5:1 mixture of $CDCl_3$/tBuOH at 298 K.

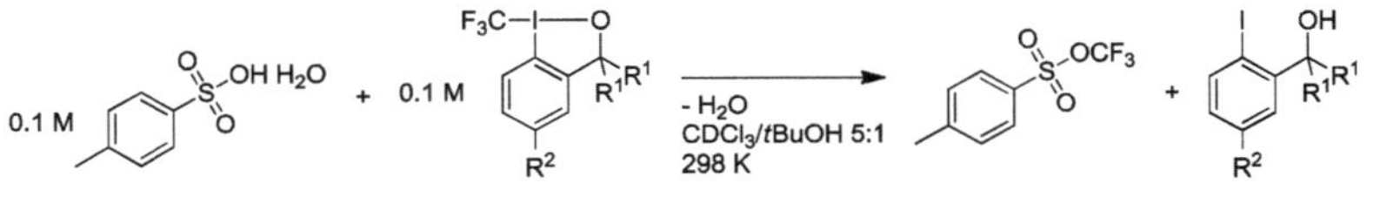

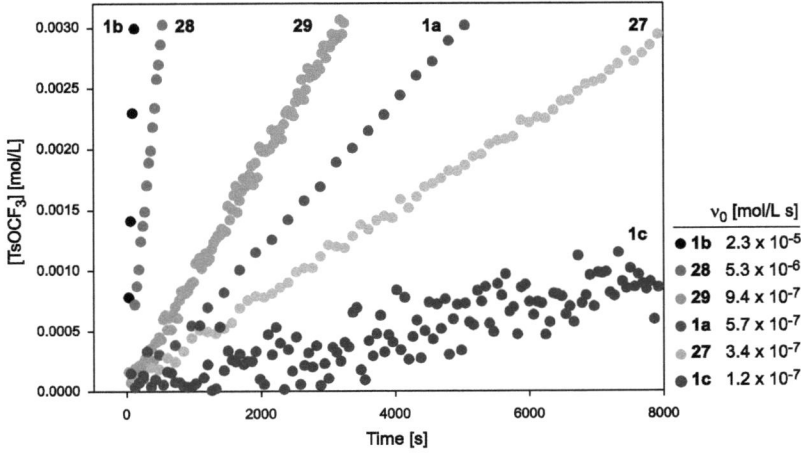

Scheme 16. Trifluoromethylation of toluenesulfonic acid monohydrate with reagents **1a-c** and **27-29**.

Figure 10. Profile for the trifluoromethylation of *para*-toluenesulfonic acid monohydrate (0.1 M) with trifluoromethylating agent **1a-c** or **27-29** (0.1 M) monitored by ^{19}F NMR and corresponding initial rates constants.

The reaction occurs smoothly at room temperature, displays a clean second order kinetics,[57b] and the results of these studies are shown in Figure 10. Compound **1b** is the most reactive compound in the electrophilic trifluoromethylation of *para*-toluenesulfonic acid ($v_0 = 2.3 \: 10^{-5}$ mol/Ls). Among the alcohol derived reagents, **28** reacts fastest ($v_0 = 5.3 \: 10^{-6}$ mol/Ls), whereas **27** ($v_0 = 3.4 \: 10^{-7}$ mol/Ls) is approximately 70 times slower than **1b** and is in the range of reagent **1a**. ($v_0 = 5.7 \: 10^{-7}$ mol/Ls). No strongly convincing correlations between X-ray structural parameters and reactivity were found and the data represent a qualitative guide, since initial rate runs were not repeated and therefore incur a certain degree of unreliability. Nevertheless, it seems that reagents containing a weakened I-O bond tend to react faster towards sulfonic acids. The assessment and comparison of the structure and rate data were complicated by strong packing effects, especially strong intermolecular interactions between the alkoxide groups and iodine(III) centers of neighboring molecules in the crystal.

2.5 Conclusion and Outlook

Several five- and a six-membered heterocyclic 1-chloro-λ^3-iodanes, including two cationic species were synthesized, as well as three new trifluoromethylating agents. The X-ray structures were compared and in a reactivity study on the trifluoromethylation of *para*-toluenesulfonic acid the structural features were correlated to the relative reactivities of the trifluoromethylation agents toward a standard substrate. Although the data should be taken as a qualitative guide, two vague trends are observed: it seems that reagents with shorter F_3C-I bond and longer I-O distances show an enhanced reactivity toward the sulfonic acid. Initial investigations suggested that 1-(trifluoromethyl)-λ^3-iodanes are largely unreactive toward hard nucleophiles (e. g. sulfonic acids) and it was found later that the presence of a strong acid is crucial for the reaction to take place, since trifluoromethylation of sodium, potassium or ammonium toluenesulfonate failed.[57b] It was mainly the work of R. Koller in collaboration with mechanistic investigations of J. M. Welch that proved that under acidic conditions the I-O bond is elongated by protonation of the reagent oxygen.[37] This observation is in agreement with the above discussed results, that 1-(trifluoromethyl)-λ^3-iodanes with long I-O bonds tend to react faster with sulfonic acids. Reagent activation opened the door to the trifluoromethylation of a variety of new substrates such as sulfonic acids,[57b] alcohols,[57c] and nitrogen centers as described in the next chapter. The application range was thus significantly expanded and the utility of reagent backbone tuning was set aside.

3 Direct Trifluoromethylation of Organonitrogen Compounds

3.1 Introduction

Throughout the last five years, a remarkable renewal of the electrophilic perfluoroalkylation chemistry has occurred, with particular emphasis on the trifluoromethyl group. The invention of new reagents with broad application range, as well as improvement in the preparation of already existing ones, has stimulated an impressive number of new methods for the electrophilic perfluoroalkylation.[36] However, despite the recent progress in the development of electrophilic trifluoromethylating agents,[3, 36] the formation of an N-CF$_3$ bond by an electrophilic trifluoromethylation reaction remains, with one exception (*vide infra*), unknown. In fact, the construction of such an entity is a particular challenge and therefore *N*-trifluoromethylated compounds are extremely rare and scarcely studied. This is revealed by an extended database search; less than ten *N*-trifluoromethylated compounds have been tested as biologically active compounds in humans,[88] and the CCDC contains only 54 crystallographically characterized NCF$_3$ derivatives,[89] most of them being perfluorinated alkyl amines. In view of the strongly electron-withdrawing nature of a trifluoromethyl group, trifluoromethylamines are anticipated to be much less basic and less nucleophilic than the corresponding methylamines.[90] Thus, the physical, chemical, and/or biological properties of trifluoromethylamines should be remarkably different from those of corresponding methylamines.[12b] In medicinal chemistry a single study undertaken in the group of Asahina the alkyl substituents on the 1-quinolone nitrogen atom of Norfloxacin and Ciprofloxacin, two important chemotherapeutic antibacterial agents, were replaced by a CF$_3$ group. The trifluoromethyl group exerts a comparable effect to that of a simple methyl group with respect to the antibacterial properties.[91] The CF$_3$ substituent was introduced by oxidative desulfurization-fluorination of the corresponding methyl dithiocarbamates, first described by Hiyama (Scheme 17).[12a, 12b] This method is still the most frequently used approach to access the NCF$_3$ unit, despite the fact that HF solutions, up to eighty equivalent F$^-$ for electron withdrawing substituents are needed,

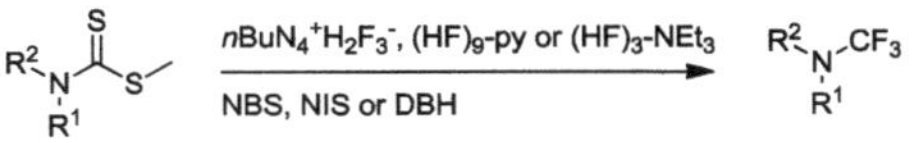

$$R^2\text{-N}(R^1)\text{-C(=S)-S-} \xrightarrow[\text{NBS, NIS or DBH}]{n\text{BuN}_4{}^+\text{H}_2\text{F}_3{}^-,\ (\text{HF})_9\text{-py or }(\text{HF})_3\text{-NEt}_3} R^2\text{-N}(R^1)\text{-CF}_3$$

Scheme 17. Oxidative desulfurization-fluorination sequence reported by Hiyama.[12a]

Other relatively mild methods for the construction of an NCF$_3$ unit by interconversion of a suitable functional group are fluorination of *N*-formamides,[92] thiuram sulphides,[93] isocyanates,[94] and trichloromethylamines,[95] the reaction of secondary amines with CBr_2F_2 and tetrakis(dimethylamino)ethylene,[96] the electrochemical fluorination of alkylamines,[97] or the fluorination of dithiocarbamates using BrF_3.[98]

Amines, anilines, and pyridines can be trifluoromethylated under very mild conditions by an *in situ* generated and thermally unstable O-(trifluoromethyl)-dibenzofuranium salt **37** following a photoirradiation protocol at −100 °C as described by Umemoto and shown in Scheme 18.[50] However, due to the inherent shortcomings of the CF_3 source employed, this methodology is not likely to replace corresponding functional group interconversions as a general method for the introduction of a CF_3 group.

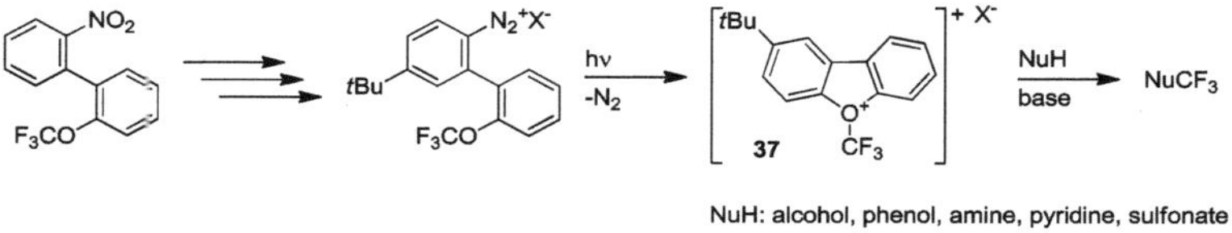

NuH: alcohol, phenol, amine, pyridine, sulfonate

Scheme 18. Electrophilic trifluoromethylation of various nucleophiles including amines and pyridines.[50]

As already noted in the previous chapters, readily accessible trifluoromethylating reagents based on hypervalent iodine react with a number of C-, S-, P-, and O-centered nucleophiles. Despite the supposed soft nature of these reagents, the substrate scope was extended to hard nucleophiles such as alcohols[57c] and sulfonic acids[57b] by activation with Lewis or Brønsted acids, respectively. A similar strategy was envisioned for the trifluoromethylation of nitrogen centers.

3.2 A Ritter-Type Reaction

3.2.1 Results

An investigation of the direct electrophilic trifluoromethylation of heteroarenes using reagent **1a** showed, in the case of nitrogen heterocycles, a pronounced tendency for the incorporation of the trifluoromethyl group at the position adjacent to nitrogen.[54c] This observation is in good agreement with early results on the trifluoromethylation on tryptophan sidechains.[55a, 99] Therefore, we were interested to see in which position trifluoromethylation would occur when the position α to the nitrogen is blocked by a substituent, as in the case of 3,5-diphenylpyrazole. We were surprised to find that under acid-catalyzed conditions in acetonitrile the main product formed was the result of a novel Ritter-type reaction in which a new N-CF_3 bond is formed leading to an N-trifluoromethylated amidine. Subsequent substrate screening showed that azoles such as benzotriazole, indazole and pyrazoles also undergo this reaction (*vide infra*). Benzotriazole (**38**) was then chosen as model substrate to optimize the reaction conditions, and the results of the screening are shown in Table 5.

Table 5. Formation of **39** under various conditions.

Entry[a]	38/1a	Acid	mol %	T [°C]	Conv.	39 Yield[b] [%]	40a Yield[b] [%]	41 Yield[c] [%]
1	1/2	HNTf$_2$	10	80	not det.	52	6	22 (44)
2	1/2	-	-	80	not det.	47	4	11 (22)
3	1/2	HNTf$_2$	10	60	not det.	60	7	25 (51)
4[d]	1/2	HNTf$_2$	5	40	not det.	44	9	18 (36)
5	1/1.5	HNTf$_2$	10	60	not det.	65	5	19 (28)
6	1/1	HNTf$_2$	10	60	99	55	5	11 (11)
7	3/1	HNTf$_2$	10	60	99	64	16	3
8	2/1	HNTf$_2$	10	60	99	67	11	5
9	1.5/1	HNTf$_2$	5	60	69	44	6	8
10[e]	1.5/1	HNTf$_2$	5	60	99	66	8	10
11	1.5/1	HNTf$_2$	10	60	99	68	7	8
12	1.5/1	HNTf$_2$	15	60	quant	70	8	8
13[e]	1.5/1	(CF$_3$)$_3$COH	10	60	quant	68	9	8
14	1.5/1	TFA	10	60	quant	60	9	8

[a] Reaction conditions: **1a** and benzotriazole (**38**) in CH$_3$CN were stirred after addition of acid (0.1 M in CH$_2$Cl$_2$) at given temperature for 3.5 h; [b] Yields calculated on the basis of ^{19}F NMR integration using C$_6$H$_5$CF$_3$ as internal standard; [c] Yields calculated on the basis of ^{19}F NMR integration using C$_6$H$_5$CF$_3$ as internal standard. Yields based on **38** are given in bracket; [d] yield after 67 h; [e] yield after full conversion (1 day).

The newly described product is formed in acetonitrile at 80 °C from benzotriazole and **1a**. However, under these conditions significant formation of HCF$_3$ (δ_F = –79.9 ppm, d, $^2J_{F,H}$ = 80 Hz) was observed by ^{19}F NMR spectroscopy, indicating reagent decomposition. The reaction temperature can be lowered to 60 °C and product yields maintained by the addition of a Brønsted acid catalyst. The best results in terms of efficiency and yield are observed when HNTf$_2$, a strong Brønsted acid with an "innocent" conjugate base, [100] is used as catalyst (Entry 11, 13, 14). Other Brønsted acids are also effective catalysts, although the reaction rate is considerably lower and lead to slightly lower yields. This is the case for (CF$_3$)$_3$COH, presumably due to its weaker acidity (pKa 5.4[101] vs 1.7[102] of HNTf$_2$) and for TFA (pKa 0.5[103]). Further reduction of the temperature to 40 °C leads to a very sluggish reaction which is not completed within two days (Entry 4). Although, as illustrated in Entry 9-12, the reaction can be accelerated by higher catalyst loadings, only a minor effect on the yield is observed. The ratio between substrate and reagent impacts the formation of the two side products **40a** and **41** (Entry 3, 5-8, 11); an excess of benzotriazole leads to the enhanced formation of direct N-trifluoromethylated benzotriazole **40a**, whereas using more **1a** favors the formation of **41**, corresponding to a Ritter-type reaction with the alcohol by-product derived from the reagent acting as nucleophile. When reagent **1a** is used as limiting species together with a slight excess of benzotriazole a good balance between side product suppression and yield optimization is observed.

The optimized conditions where then applied to other azoles, and the results of the substrate screening are shown in Table 6.

Table 6. Substrate screening under standard conditions for Ritter-type reaction.

Entry[a]	Substrate		Product		NMR Yield[b] [%]	Yield [%]
1	38		39		68	63
2[c]	38		50		60	37
3[d]	38		51		48	36
4[e]	38		52		19	14
5[f]	38		53		21	7
6	42		54a		57	47
7[g]	43		55		47[h]	42[h]
8[c][g]	43		56		37	35

continued on the next page

Direct Trifluoromethylation of Organonitrogen Compounds

Entry[a]	Substrate		Product		NMR Yield[b] [%]	Yield [%]
9[c]	44	tBu-pyrazole NH	57	tBu-pyrazole N-N=C(CF$_3$)	58	not det.[i]
10	45	Mes-pyrazole NH	58	Mes-pyrazole N-N=C(CF$_3$)	54	51
11[j]	46	tBu,tBu-pyrazole NH	59	tBu,tBu-pyrazole N-N=C(CF$_3$)	52	47
12	47	Ph,Ph-pyrazole NH	60	Ph,Ph-pyrazole N-N=C(CF$_3$)	59	47
13	48	Me-pyrazole NH	61	Me-pyrazole N-N=C(CF$_3$)	62	53
14[k]	49	EtO$_2$C-pyrazole NH	62	EtO$_2$C-pyrazole N-N=C(CF$_3$)	47	38

[a] Reaction conditions: **1a** and azole (1.5 equiv) in CH$_3$CN were stirred. After addition of HNTf$_2$ (10 mol-%, 0.1 M in CH$_2$Cl$_2$) the mixture was heated to 60 °C for 3.5 h; [b] Yields calculated on the basis of ^{19}F NMR integration using C$_6$H$_5$CF$_3$ as internal standard; [c] C$_2$H$_5$CN instead of CH$_3$CN; [d] /PrCN instead of CH$_3$CN; [e] PhCN instead of CH$_3$CN; [f] BnCN instead of CH$_3$CN; [g] Reaction time elongated to 6 h; [h] N^2-isomer was formed in 9% NMR yield, 3% isolated; [i] not isolated in pure form; [j] Reaction time elongated to 16 h; [k] Due to separation problems: addition of Burgess-reagent[104] (1.5 equiv in CH$_2$Cl$_2$) after completed reaction, the mixture stirred for additional 30 min at 60 °C.

All azoles tested are converted to their corresponding *N*-trifluoromethylated amidines in moderate to good yields and can easily be separated from the much more polar starting azoles by simple flash column chromatography. The Ritter-product **62** of the reaction of ethyl 4-pyrazolecarboxylate (**49**), reagent **1a** and acetonitrile is not stable under the reaction conditions and longer reaction times lead to lower yield. After full conversion, the desired product is formed in 68% yield, but, unfortunately, is not separable from the reagent by-product. After dehydrogenation of the latter with the Burgess-reagent, the desired product can be isolated in 38% yield in pure form after flash column chromatography. Instead of acetonitrile, the above described reaction can also be carried out in propionitrile, and the corresponding *N*-trifluoromethylated imine **50** is obtained in slightly lower yields compared to the analogous reaction in CH_3CN. For more sterically demanding and conjugated nitriles such as iPrCN, $PhCH_2CN$ or PhCN significant decreases of yields relative to CH_3CN are observed. Under standard conditions 1H-indazole (**42**) is converted to the N^2 trifluoromethylamidine substituted heterocycle **54a** in 57% NMR yield, with the formation of only 5% of the corresponding N^1-substituted indazole **54b**. The former compound can be isomerized to the N^1-derivative within 24 h upon heating to 70 °C in acetonitrile in the presence of 10 mol-% $HNTf_2$ in 79% NMR yield.

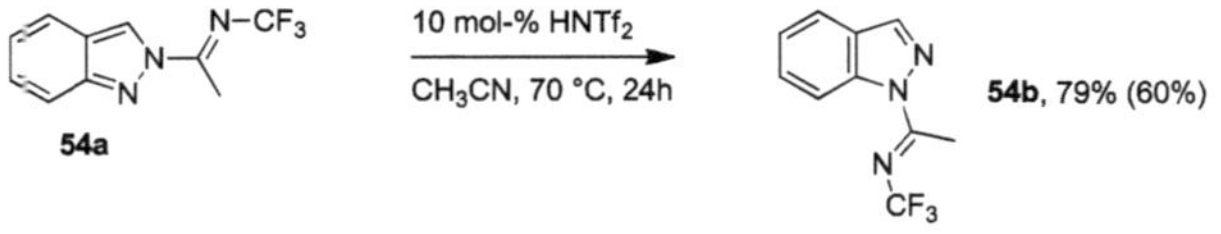

Scheme 19. Isomerization of N^2-substituted indazole **54a** to N^1-substituted derivative **54b**.

These experiments show that the N^1- and N^2-subsituted compounds correspond to the thermodynamic and kinetic products, respectively. A comparable isomerization reaction concerning N-(N',N'-dialkylaminomethyl)benzotriazoles has been published by Katritzky to take place via a dissociation recombination mechanism involving an iminium cation by cleavage of the C-N-bond proved by cross-over experiments.[105] Therefore, it seems reasonable to assume that the tautomerism of the indazole derivatives also proceeds *via* a cationotropic process involving the dissociation and recombination of a trifluoromethyl nitrilium ion.

3.2.2 Structure Determination

Since the newly prepared compounds contain an uncommon N-(trifluoromethyl)-amidine unit, a thorough structural investigation was carried out. Representations of the crystal structures of the reaction products of benzotriazole, indazole (both isomers) and 3,5-diphenylpyrazole are shown in Figure 11 and selected bond lengths and angles are collected in Table 7. In all structures the alkyl subsitutents in the newly formed trifluoromethylamidine group are Z-configurated, and the exocyclic single bond has an *s-trans* conformation. While the imidoyl group is twisted out of the pyrazole and slightly from the benzotriazole plane, as indicated by the N-N-C-N torsion angles, the indazole derivates show almost perfect planarity, as there are no 1,5 or 1,6-repulsions possible. The newly formed N-CF_3 and N^1-$C^{imidoyl}$ bonds are nearly equal both in the range of short single bonds i.e. 1.4 Å. This indicates significant conjugation despite the twist. The N-C-C bond angle in the imidoyl unit is enlarged and deviates up to 10° from the expected ideal geometry.

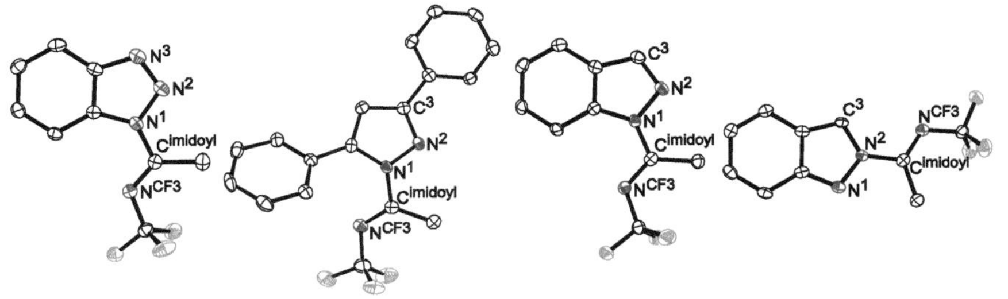

Figure 11. ORTEP drawings of X-ray the structures of **39, 60, 54b, 54a** (from left to right). Hydrogen atoms are omitted for clarity, thermal ellipsoids set to 50% probability.

Table 7. Compilation of selected bond lengths, bond-, and torsion angles.

Bond lengths [Å]	39	60	54b	54a
N^1-N^2	1.383(2)	1.374(4)	1.390(2)	1.363(2)
N^2-N^3/C^3	1.283(3)	1.387(6)	1.302(2)	1.360(3)
N^1/N^2-$C^{imidoyl}$	1.399(2)	1.403(5)	1.383(2)	1.408(3)
$C^{imidoyl}$-N^{CF3}	1.274(2)	1.273(5)	1.283(2)	1.276(3)
C-CH_3	1.496(3)	1.491(6)	1.499(2)	1.488(3)
N-CF_3	1.399(2)	1.395(5)	1.387(2)	1.389(3)
Bond angles [°]				
N^1-N^2-N^3/C^3	109.0(1)	111.6(3)	105.69(15)	113.52(17)
N^{CF3}-C-N^1/N^2	115.1(1)	115.6(4)	115.88(16)	115.38(18)
N^{CF3}-C-CH_3	130.2(1)	129.3(4)	128.52(17)	129.9(2)
Torsion angles [°]				
N-N-C-N	−172.9(1)	−147.9(4)	176.71(15)	−178.06(17)

In addition to studies of the new compounds in the solid-state, the structures in solution have been established by multinuclear NMR methods. As representative examples the $^{19}F^1H$ HOESY, $^1H^{15}N$ and $^{19}F^{15}N$ HMQC spectra of **39** are shown in Figure 12. The $^{19}F^1H$ HOESY spectrum shows a contact between the fluorine atoms and the protons of the CH_3-group derived from acetonitrile, thus confirming the *E*-configuration of the newly formed group. In the $^1H^{15}N$ correlation spectrum, two cross-peaks between the methyl resonance (δ_H = 3.12 ppm) and nitrogen are obtained. The nitrogen resonances can be unambiguously assigned by $^{19}F^{15}N$ correlation, since only one correlation between the trifluoromethyl resonance (δ_F = −53.5 ppm) and the imido resonance (δ_N = 254.5 ppm) is observed. Therefore, the resonances at 245.3 ppm and 254.5 ppm can be assigned to azole and imido nitrogen nuclei, respectively. In addition, since the spectra were measured without broadband decoupling, this method allows the determination of a $^2J_{F,N}$ coupling of about 20 Hz. It has to be noted that values for $^2J_{F,N}$ are scarcely found in the literature.

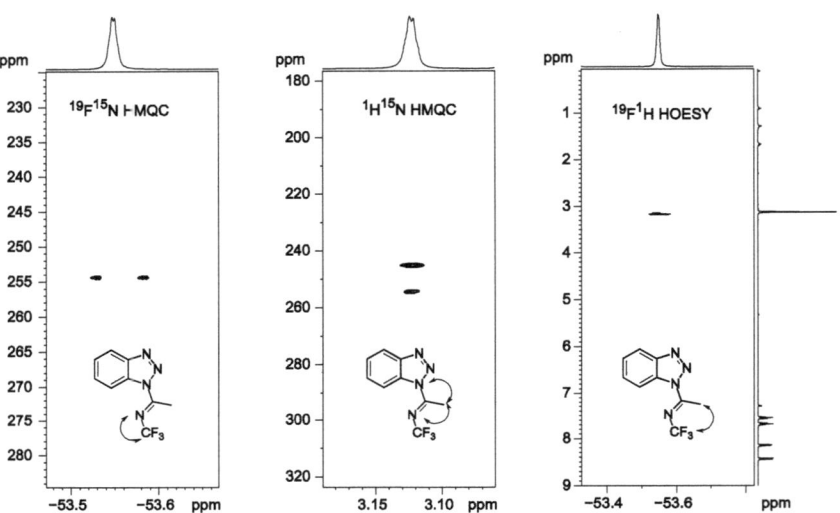

Figure 12. $^1H^{15}N$ (left), $^{19}F^{15}N$ HMQC (middle) and $^{19}F^1H$ HOESY (right) spectra of **39**.

3.3 Mechanistic Investigations

To gain a deeper insight into the actual processes, the reaction was monitored by ^{19}F NMR spectroscopy. The reaction profile for benzotriazole (0.15 M), reagent (0.1 M) and HNTf$_2$ (6.8 mM) in CD$_3$CN is shown in Figure 13. The rate of reagent consumption and product formation appears to be largely constant. The small deviations from linearity observed can be explained by side product formation. These observations imply that a constant concentration of the active species is present over the course of the reaction. Acetonitrile is present in large excess and therefore the concentration is essentially constant. This is true also for the Brønsted acid present in catalytic amounts. Over the course of the reaction, the reagent flourine resonance shifts by ca. 20 ppm to higher frequency. Furthermore, comparison of the ^{13}C NMR spectra of the reagent and the reagent protonated by one equivalent of acid showed that all aromatic carbon resonances are shifted and the carbon fluorine coupling constant of the trifluoromethyl group is decreased ($J_{C,F}$ = 321 Hz vs $J_{C,F}$ = 396 Hz). This implies, together with previous observations concerning the trifluoromethylation of THF[57e] and toluenesulfonic acid,[37, 57b] that the reaction involves protonation of the reagent **1a**. On the basis of this study the following reaction mechanism shown in Scheme 20 is proposed.

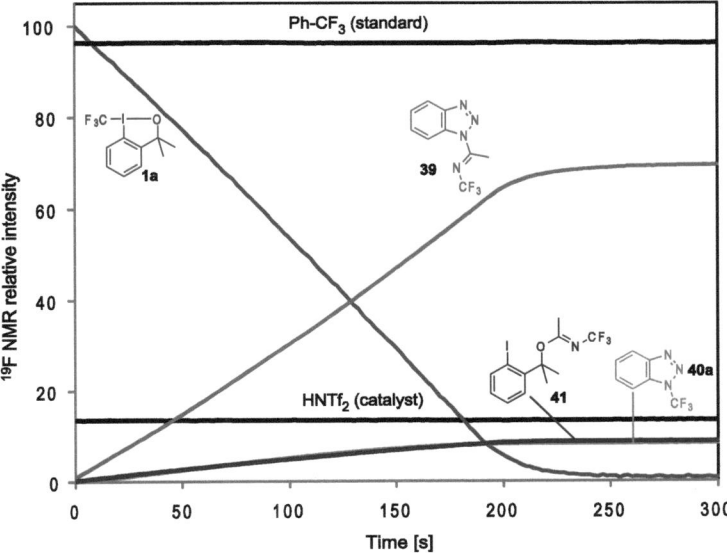

Figure 13. Profile for the reaction of 0.1 M **1a** with 0.15 M **39** and 6.8 mM HNTf$_2$ in CD$_3$CN as monitored by ^{19}F NMR, C$_6$H$_5$CF$_3$ internal standard. Formation of around 15% HCF$_3$ is not included in the figure.

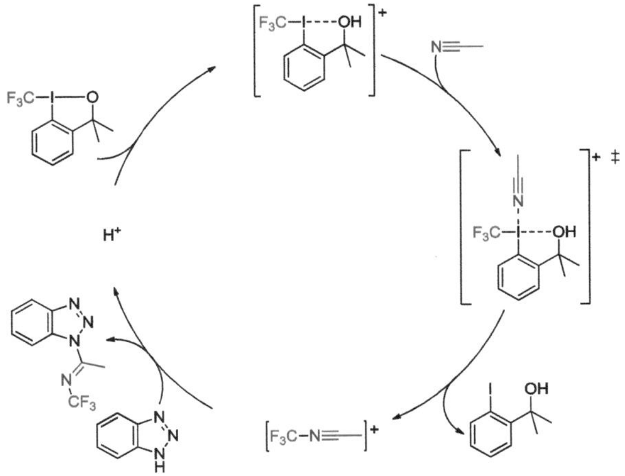

Scheme 20. Proposed reaction mechanism for acid-catalyzed Ritter-type reaction of benzotriazole.

Direct Trifluoromethylation of Organonitrogen Compounds

In the first step, the reagent is protonated, weakening the I-O bond and thus making the iodine atom more electrophilic. In the transition state, acetonitrile attacks the protonated form of reagent **1a**, liberating reduced reagent and an N-trifluoromethylated nitrilium ion. The assumption that acetonitrile interacts with the hypervalent iodine center in the rate determining step seems reasonable, since the formation of complexes of I(III) with nitrogen bases such as acetonitrile and pyridine have been reported.[106] However, PGSE NMR measurements show that the protonated reagent and acetonitrile diffuse independently, confirming that the one to one transition state stoichiometry implied by the kinetic measurements does not correspond to an intermediate. The nitrilium ion formed in the rate-determining step is then trapped by benzotriazole forming the product and releasing a proton that reenters the catalytic cycle.

The trifluoromethylated nitrilium ion can be formed by a reductive elimination or an S_N2 pathway which cannot be experimentally distinguished. Therefore, in a collaboration with PD H. P. Lüthi this reaction step was investigated by DFT studies (B3LYP/aug-cc-pVDZ-pp).[107] These calculations suggest a slightly different mechanistic situation but fundamentally yield the same results (Scheme 21).

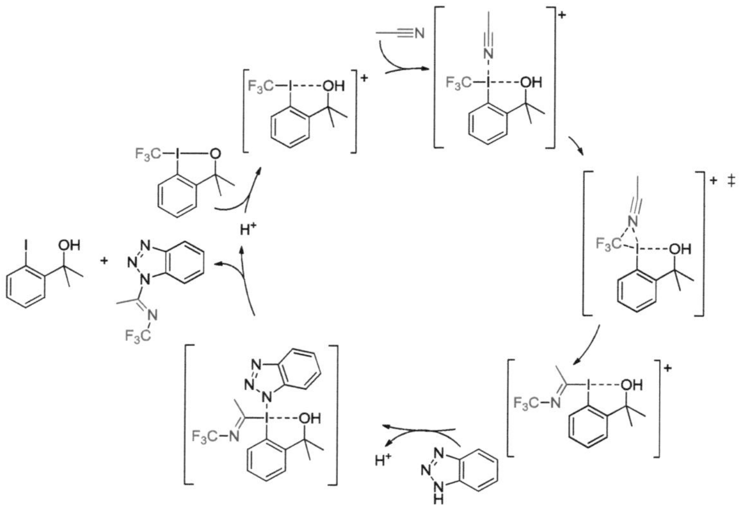

Scheme 21. Mechanism for Ritter-type reaction based on DFT calculations. Program: Gaussian 09, hybridfunctional: B3LYP, basis-set: aug-cc-pVDZ, ECP: Stuttgart-Koeln MCDHF.[107a]

The first two steps are identical to the above discussed mechanism: reagent activation and coordination of acetonitrile. In the transition state, the trifluoromethyl group and the nitrile nitrogen are rearranged to form an iodonium species with two carbon ligands. A deprotonated benzotriazole coordinates to the newly formed intermediate. After a reductive elimination, product and reduced reagent are formed, releasing a proton that reenters the catalytic cycle. The activation barrier for the transition state was calculated to be 39.4 kcal/mol, and the newly formed intermediate is stabilized by 2.1 kcal/mol. An activation energy of only 3.3 kcal/mol is needed for the reductive elimination with coordinated deprotonated benzotriazole. This might explain why direct N-trifluoromethylated benzotriazole is only formed as a side product. Firstly, the coordination sites of iodine are blocked by acetonitrile and secondly, benzotriazole has to be deprotonated to react further. Since the mechanism proposed in Scheme 21 has the same transition state as nitrilium formation *via* a reductive elimination (Scheme 20), and since the activation barrier is rather large, such a pathway is rather unlikely.

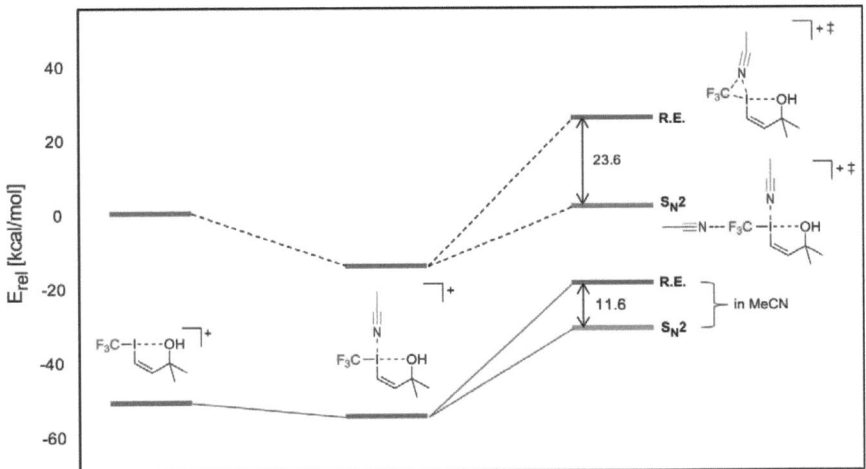

Figure 14. Energy diagram comparing the two competing reaction mechanism in gasphase (upper) and in acetonitrile (lower). Program: Gaussian 09, hybridfunctional: B3LYP, basis: aug-cc-pVDZ-pp, ECP: Stuttgart-Koeln MCDHF.[107b]

Further calculations on a simplified reagent substitute DMTI (3,3-dimethyl-1-(trifluoromethyl)-λ^3-2-iodoxol in gasphase as well as in solution were performed, and the results are shown in Figure 14. A stabilization of 15 kcal/mol (gasphase) and 3.5 kcal/mol (solution), for the coordination of acetonitrile to the hypervalent iodine is calculated. The calculations support a nucleophilic reaction with a second acetonitrile to form a nitrilium ion rather than a reductive elimination (R.E.) process. The activation energies for an S_N2 reaction are calculated to be 17.0 kcal/mol (gasphase) and 24.3

kcal/mol (solution), resulting in activation energies which are 23.6 kcal/mol (gasphase) and 11.6 kcal/mol (solution) smaller than those calculated for the reductive elimination process.

3.4 Direct *N*-Trifluoromethylation of Benzophenone Imine

The reaction of benzophenone imine (**63**) and reagent **1a** under the same reaction conditions as for azoles was also tested. Interestingly, benzophenone imine did not react with **1a** under acid-catalyzed conditions in acetonitrile to the corresponding Ritter-type product. Instead in a slow reaction (22% conversion after 41 h) the imine was *N*-trifluoromethylated directly (Table 8, Entry 1). Table 8 shows the screening results for the optimization of the direct *N*-trifluoromethylation of benzophenone imine.

Table 8. Condition screening for direct *N*-trifluoromethylation of benzophenone imine.

Entry[a]	Acid/Additive	mol%	Solvent	Time [h]	Yield[b] [%]
1	HNTf$_2$	10	CH$_3$CN	41	14[c]
2	HNTf$_2$	100	CH$_3$CN	3	21
3	TFA	100	CH$_3$CN	0	3
4	TMSCl	100	CH$_3$CN	15	39
5	TMSNTf$_2$	100	CH$_3$CN	3.5	3
6	TMSNTf$_2$	10	CH$_3$CN	18	25
7	(TMS)$_3$SiCl	100	CH$_3$CN	20	66
8	(TMS)$_3$SiCl	100	CH$_3$CN	24	54
9	(TMS)$_3$SiCl	100	CDCl$_3$	48	40
10	(TMS)$_3$SiCl	100	DCE	16	60
11	(TMS)$_3$SiCl	100	DMF	16	46
12	(TMS)$_3$SiCl	100	toluene	24	8

[a] Reaction conditions: reagent **1a** and benzophenone imine (1.5 equiv) in given solvent after addition of additive or acid at 60 °C; [b] Yields calculated based on integration of ^{19}F NMR signals using C$_6$H$_5$CF$_3$ as internal standard; [c] Conversion 22%.

Neither the addition of an equivalent of acid, HNTf$_2$ (Entry 2) or TFA (Entry 3), nor the use of TMSNTf$_2$ (Entry 5/6) improved the reaction. However, first improvements in yield were obtained, when one equivalent of TMSCl (Entry 4) under complete exclusion of air and moisture was added to the reaction mixture. Even better results were obtained after the addition of bulky (TMS)$_3$SiCl, but changing the solvent decreased the product yield. Furthermore, it was observed that the product is not stable under the reaction conditions (compare Entry 7 to 8). Using the optimized conditions from Entry 7, *N*-trifluoromethylated benzophenone imine could be isolated after flash column chromatography in reasonable purity, approx. 95% by NMR. However, the product decomposes in solution as well as in the solid state even at low temperatures (−18 °C) and was therefore characterized only by NMR and HRMS.

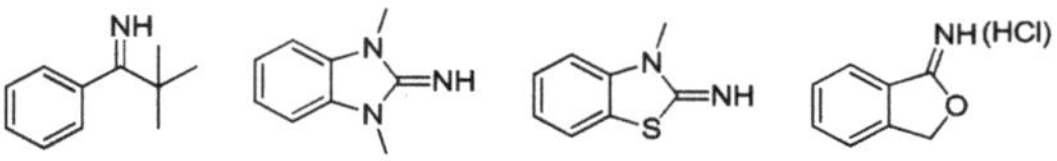

Figure 15. Benzophenone imine related substrates tested unsuccessfully in direct N-trifluoromethylation reaction utilizing (TMS)$_3$SiCl.

Related substrates, shown in Figure 15 were not converted to the corresponding N-trifluoromethylated compounds. It cannot be ruled out that the products are actually formed, but are simply not stable enough under the reaction conditions.[108]

3.5 Direct N-Trifluoromethylation of Heterocycles

3.5.1 Reaction Optimization

In Section 3.2.1 it was briefly mentioned that N-trifluoromethylbenzotriazole **40a** is among the side products of the Ritter-type reaction. In 2000, Yagupolskii published the synthesis of N-trihalomethyl derivatives of benzimidazole, benzotriazole and indazole. In this report, **40a** is prepared by functional group interconversion from its trichloro derivative and is the only report on the synthesis of nitrogen trifluoromethylated azoles to date.[109] Therefore the observation of **40a** among the side products of the Ritter-type reaction is crucial, indicating that the direct N-trifluoromethylation of azoles, a potentially useful and desirable, but still essentially unknown reaction, is indeed possible. A closer look at Table 5 reveals that the extent to which compound **40a** is formed strongly depends on the substrate-to-reagent ratio with an excess of substrate enhancing the relative amount of **40a** formed. Furthermore, a quick substrate screening in a solvent that was not able to undergo the Ritter-type reaction (DCE or 1,1',2,2'-tetrachloroethane) showed that benzotriazole as well as pyrazoles could be directly N-trifluoromethylated in 41% yield (Table **9**, Entry 2). When the reaction was carried out in highly concentrated CS$_2$ solutions, up to 60% yields could be achieved (Entry 3). Unfortunately, this solvent proved to be a poor choice for substrates other than benzotriazole. Taking up previous findings concerning the reaction of silyl enol ethers[7, 37] and silylated phosphines [56a] the reaction was also carried out with silylated benzotriazole **65**. Depending on the choice of silylated versus protonated benzotriazole and acid catalyst different temperatures are needed to obtain optimal product yields.

Table 9. Condition screening for direct *N*-trifluoromethylation of benzotriazole to give product **40a**.

R = H **38**
R = TMS **65**

Entry	[equiv][a]	R	Acid	Mol-%	Solvent	T [°C]	Conc.[b] [M]	Yield[c] [%]
1	3	H	HNTf$_2$	10	CH$_3$CN	60	0.1	16
2	1.5	H	HNTf$_2$	10	DCE	60	0.1	41
3	1.5	H	HNTf$_2$	10	CS$_2$	60	1.5	58
4	1.5	H	HNTf$_2$	10	DCM	35	0.1	6
5	1.5	H	HNTf$_2$	10	DCM	35	1.5	16
6	1.5	TMS	(Tf$_2$CH)$_2$CH$_2$	10	DCM	r.t.	0.1	45
7	1.5	TMS	(Tf$_2$CH)$_2$CH$_2$	5	DCM	r.t.	0.1	42
8	1.5	TMS	(Tf$_2$CH)$_2$CH$_2$	10	DCM	r.t.	0.2	51
9	1.5	TMS	(Tf$_2$CH)$_2$CH$_2$	10	DCM	r.t.	0.5	55
10	1.5	TMS	HNTf$_2$	10	DCM	35	0.1	42
11	1.5	TMS	HNTf$_2$	10	DCM	35	0.2	50-59
12	3	TMS	HNTf$_2$	10	DCM	35	0.2	80
13	1.5	TMS	HNTf$_2$	10	DCM	35	0.5	70-73
14	1.5	TMS	HNTf$_2$	10	DCM	35	1.0	79
15	1.5	TMS	HNTf$_2$	10	DCM	35	1.5	81
16	1.5	TMS	HNTf$_2$	7.5	DCM	35	1.5	80
17	1.5	TMS	HNTf$_2$	5	DCM	35	1.5	76
18	1.1	TMS	HNTf$_2$	10	DCM	35	1.5	73
19	1.1	TMS	HNTf$_2$	12	DCM	35	1.5	87[d]

[a] Based on limiting reagent **1a**; [b] concentration based on reagent; [c] yields calculated based on integration of ^{19}F NMR signals using C$_6$H$_5$CF$_3$ as internal standard; [d] addition of LiNTf$_2$ before the addition of reagent.

In general, higher temperatures are necessary for a successful transformation of 1*H*-benzotriazole (**38**). When **65** was used the temperature could be lowered to 35 °C or in combination with (Tf$_2$CH)$_2$CH$_2$ as catalyst even to room temperature. In both cases, better yields were observed when working at high concentrations. This effect was even more pronounced for the reactions with HNTf$_2$ as catalyst. A reasonable explanation for this finding is that at higher concentrations the trifluoromethylation reaction is accelerated, whereas the reaction rates of the side reactions, mainly catalyst and reagent decomposition as well as TMSF formation, are not as strongly affected and therefore after reaction completion the acid catalyst is not fully decomposed. The catalyst loading has only a minor effect on product yield when working at high concentrations. Although an excess of nucleophile is beneficial in terms of product yield, using only a slight excess of substrate (1.1 equiv instead of 1.5) leads to somewhat reduced yield (Entry 18).

In order to gain a deeper insight into the actual processes involved in the reaction and to further optimize the reaction conditions, we monitored the reaction by ^{19}F NMR spectroscopy. The reaction profile for 1-TMS-benzotriazole (1.65 M), **1a** (1.5 M) in CD$_2$Cl$_2$ and internal standard (PhCF$_3$, 0.29 mmol), after the addition of 12 mol-% HNTf$_2$ is shown Figure 16.

Both reagent decay and product formation show an exponential behavior. The main product observed is N^1-trifluoromethylated benzotriazole **40a**, whereas the N^2-isomer is only formed in traces. The concentration of acid catalyst HNTf$_2$ decreases slightly over the course of the reaction, and therefore can lead, in cases of slow reaction rates, to full decomposition, causing reaction termination and low yields. Furthermore, CF$_3$H and unstable trifluoromethylated benzyl alcohol **66** are formed, both being decomposition products of **1a**. The concentration of a catalytic amount of activated reagent (by protonation or silylation) is diminished over the course of the reaction and seems to be related to TMSF formation. This correlation suggests fluoride formation, which might interfere with the reaction.

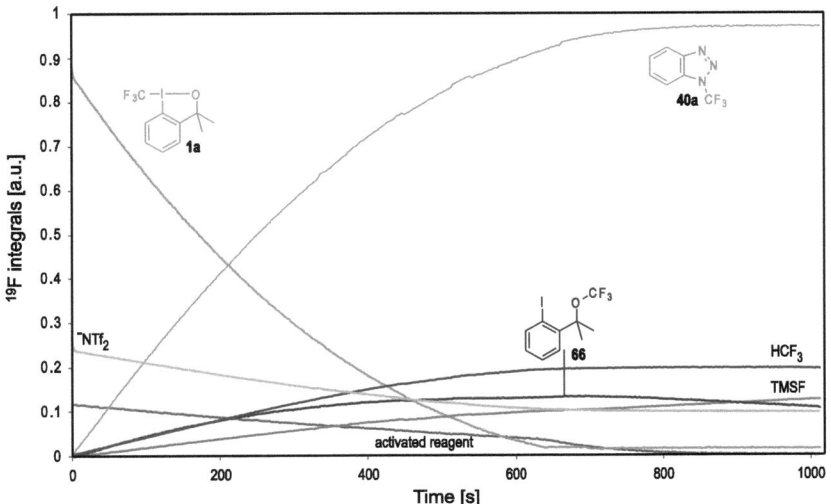

Figure 16. ^{19}F NMR reaction profile of electrophilic trifluoromethylation of **65** with 1.5 M **1a** in CD$_2$Cl$_2$ using 12 mol-% HNTf$_2$ as acid catalyst.

Therefore, an additional ^{19}F NMR experiment (Figure 17) was carried out with 5 mol-% BF$_3$-Et$_2$O, a catalyst able to remove fluoride from the reaction mixture. In this experiment N^1- and N^2-trifluoromethylated benzotriazole are observed and both products are formed at essentially identical exponential rates in an overall yield of 94%. The ratio of the two isomers is dependent on the catalyst, a result that is to be expected since alkylation reactions of heterocycles are known to be highly sensitive to reaction conditions.[105, 110] Again, an exponential decay of reagent is observed, whereas the formation of 3% trifluoromethylated benzyl alcohol **66** accounts for the only notable side product. This experiment supports the conjecture that fluoride formation interferes with the reaction and shows the beneficial effect of a potential fluoride scavenger.

However, as promising as BF$_3$ was as a catalyst in the aforementioned experiment, it was significantly less effective in combination with other silylated nitrogen heterocycles.

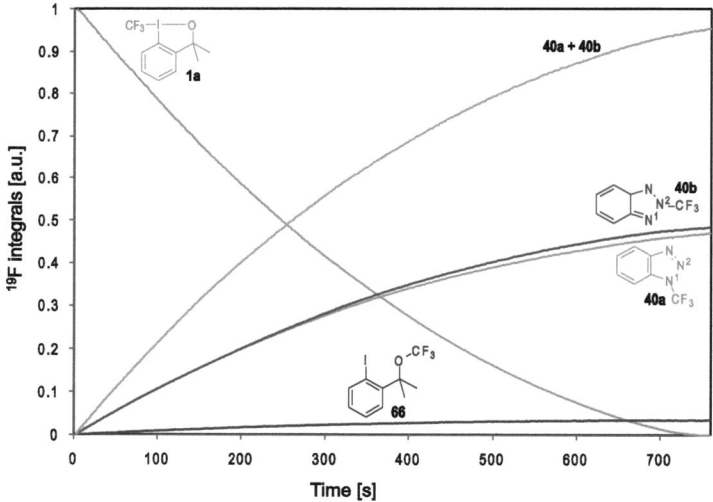

Figure 17. ^{19}F NMR reaction profile of the electrophilic trifluoromethylation of **65** with 1.5 M **1a** in CD$_2$Cl$_2$ using 5 mol-% BF$_3$-Et$_2$O as acid catalyst.

Therefore, 2 mol-% of non-Lewis acidic LiNTf$_2$, an alternative fluoride scavenger compatible with a wide range of substrates, were added to the reaction mixture using HNTf$_2$ as acid catalyst. *N*-trifluoromethylated benzotriazole was formed in excellent 87% yield, when LiNTf$_2$ was present at the onset of the reaction (Table 8, Entry 19), but had no effect on yield when added after the addition of reagent.

To circumvent the isolation of moisture sensitive silylated azoles an *in situ* silylation sequence was developed. HMDS with a catalytic amount of silica sulfuric acid (SSA) was chosen as silylating agent, since it had been reported to silylate several *N*-heterocycles in quantitative yields within short reaction times.[111] Furthermore, SSA is an efficient, inexpensive, strong, reusable, heterogenous acid catalyst that allows an easy workup by simple filtration.[112] When benzotriazole was silylated *in situ* under reflux in HMDS with a catalytic amount of SSA, followed by filtration and exchanging the solvent to DCM prior to trifluoromethylation, only a minor decrease in yield for **40a** was observed. Instead of 87% yield from isolated 1-TMS-benzotriazole (**65**) the yield dropped to 84% when the *in situ* silylation sequence was applied. To simplify the addition of hygroscopic HNTf$_2$ to moisture sensitive silylated azoles, the second reaction step of the following transformation was normally carried out in a glovebox. On a larger scale, similar yields were observed when the reaction was carried out using the standard Schlenk technique, with an HNTf$_2$ stock solution.

3.5.2 Substrate Scope

The optimized reaction conditions, as developed above, that is with increased in overall concentration in a solvent that does not undergo a Ritter-type reaction, addition of LiNTf$_2$ prior to the onset of the trifluoromethylation reaction, and use of *in situ* silylated azole, were successfully applied to a variety of different azoles, such as pyrazoles, indazoles, triazoles, tetrazole and to a certain extent benzimidazole as summarized in Table 10. The sterically demanding 3,5-di-*tert*-butyl- and 3,5-diphenylpyrazole could neither be *C*-trifluoromethylated by the standard procedure nor directly at nitrogen from the isolated silylated derivatives. Thus, **70** and **72**, pyrazoles with two bulky substitutents in 3- and 5-position, do not afford the desired products primarily due to steric reasons, since heterocycles with electron-donating groups generally give higher yields (Table 10, Entry 3-5). Pyrazoles with alkyl, aryl and alkoxycarbonyl substitutents undergo the desired reaction and various substitution patterns are tolerated. While electron rich substituents accelerate the reaction, electron deficient heterocycles show slower reaction rates, allowing for side product formation to predominate (mainly decomposition of the reagent). Typically, when working with unsymmetrically substituted substrates, isomeric product mixtures were obtained that were separated by flash column chromatography. Most pyrazoles are preferentially trifluoromethylated in N^1-position, whereas for indazoles the N^2-products are favoured (Entry 10/11). As already discussed in Section 3.4.1. the isomeric distribution for benzotriazole depends on the acid catalyst. When HNTf$_2$ is used under standard conditions, the formation of the symmetric N^2-substituted product is almost completely inhibited (Entry 14/15), while under the same conditions product mixtures are obtained for other triazoles (Entry 12/13).

Direct Trifluoromethylation of Organonitrogen Compounds

Table 10. Substrate scope for the direct *N*-trifluoromethylation of different azoles.

Entry	Substrate	Major Product	Minor Product	NMR Yield[a] [%]	Yield [%]
1	43	79		69	62
2[b]	45	80a	80b	34/28	30/25[c]
3	67	81a	81b	35/7	33[d]/-
4[b]	49	82		26	(<13)
5	68	83		69	66
6[e]	69	84		68	15
7[e]	70			-	-
8	71	85a	85b	48/20	40/12
9[e]	72			-	-
6[e]	69	84		68	15
7[e]	70			-	-

continued on the next page

Entry	Substrate	Major Product	Minor Product	NMR Yield[a] [%]	Yield [%]
8	71	85a	85b	48/20	40/12
9[e]	72			-	-
10	73	86a	86b	49/29	30/12[f]
11	42	87a	87b	68/2	39[g]/-
12	74	88a	88b	46/10	24[h]/-
13	75	89a	89b	42/11	(21/11)
14	38	40a	40b	84/2	(64/-)
15[i]	38	40b	40a	48/46	(44/24)
16[j]	76	90a	90b	42/18	(18/10)
17[e]	77			-	-
18	78	91		21/16	13/-

[a] Yields calculated based on integration of ^{19}F NMR signals using $C_6H_5CF_3$ as internal standard; [b] 14 mol-% HNTf$_2$; [c] **80b** contained 3% **80a**, 15% yield as single regioisomer after sublimation; [d] **81a** contained 5% **81b**; [e] without TMS-protection sequence; [f] **86b** contain ≤5% **86a**; [g] **87a** contained <2% **87b**; [h] **88a** contained 8% **88b**; [i] 5 mol% BF$_3$-Et$_2$O instead of LiNTf$_2$ and HNTf$_2$; [j] r.t. instead of 35 °C, and no addition of HNTf$_2$.

After slight modifications of the standard reaction procedure, reduction of reaction temperature to ambient temperature for the trifluoromethylation step and no acid catalyst addition, 5-phenyltetrazole (**76**) was *N*-trifluoromethylated in good yield (Entry 16). The *N*-trifluoromethylation of benzimidazole proved difficult. Most derivatives are largely insoluble in chlorinated solvents,[113] and for unsubstituted 1*H*-benzimidazole trifluoromethylation occurs preferentially at the position adjacent to nitrogen.[54c] Slightly even less soluble 2-methyl-benzimidazole (**77**) was tested as substrate for the direct *N*-trifluoromethylation, but no product was observed (Entry 17). The successful *N*-trifluoromethylation of 2-ethylthio-1*H*-benzo[*d*]imidazole (**78**) demonstrated that benzimidazoles can, in principle, be trifluoromethylated as well, although only in low yield (Entry 18).

3.5.3 Product Characterization

The novel methods developed allow the synthesis of a variety of *N*-trifluoromethylated azoles, which are otherwise difficult to obtain. Therefore, it is not surprising that most products described in Table 10 have never been reported before. After separation and purification by flash column chromatography the new compounds were fully characterized. Multinuclear NMR spectroscopy proved to be a valuable tool to assign the regioisomers formed during the trifluoromethylation reaction. As a representative example, the ^{19}F^{1}H HOESY and ^{19}F^{15}N HMBC correlation spectra of the major and minor isomer of **86a** and **86b** are shown in Figure 18.

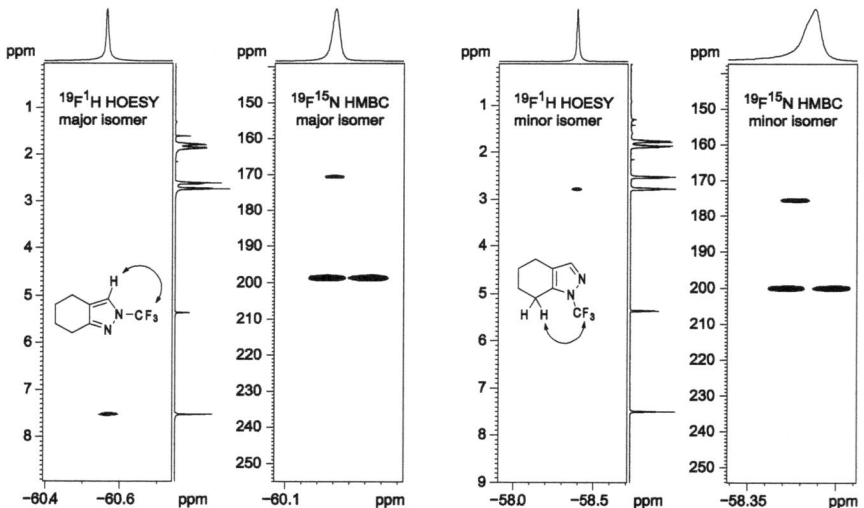

Figure 18. Left side: ^{19}F^{1}H HOESY (left) and ^{19}F^{15}N HMBC (right) spectrum of major isomer **86a**; right side: ^{19}F^{1}H HOESY (left) and ^{19}F^{15}N HMBC (right) spectrum of minor isomer **86b**.

In the ^{19}F^{1}H HOESY spectrum of the major regioisomer a signal due to the proximity of the CF$_3$ group to the proton on the heteroaromatic ring is observed, whereas the spectrum of the minor isomer shows a cross peak between the trifluoromethyl group and an aliphatic proton. The ^{19}F^{15}N HMBC correlation spectrum allows the unambiguous assignment of the ^{15}N resonance of the nitrogen bearing the CF$_3$ group due to the $^2J_{F,N}$ coupling of ca. 20 Hz.

In addition, the structures of **79, 80b, 81a, 85a, 89a** were determined by single crystal X-ray analysis and their ORTEP representations are shown in Figure 19 and selected bond lengths and angles are summarized in Table 11. As expected, the N-CF$_3$ bond lengths are slightly elongated in comparison to the products of the Ritter-type reaction. In all cases the N^2-N^1-CF$_3$ bond angle is around 10° smaller than the C^5-N^1-CF$_3$ angle, independent of the substitution pattern. Furthermore, the CF$_3$ group does not perfectly lie in the heterocyclic plane as shown by the C^3/N^3-N^2-N^1-CF$_3$ torsion angles, although N^1 shows in none of the cases pyramidalization.

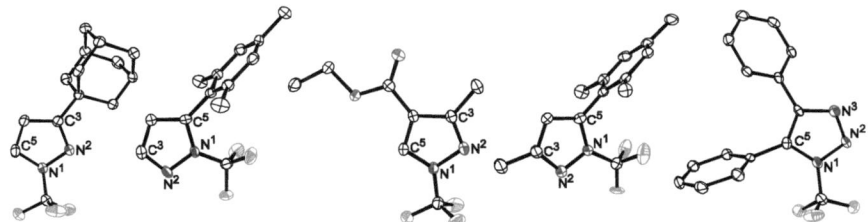

Figure 19. ORTEP drawings of the X-ray structures of **79, 80b, 81a, 85a, 89a** (from left to right). Hydrogen atoms are omitted for clarity, thermal ellipsoids set to 50% probability.

Table 11. Compilation of selected bond lengths, bond-, and torsion angles.

Bond lengths [Å]	79	80b	81a	85a	89a
N^1-N^2	1.368(2)	1.372(3)	1.376(2)	1.369(3)	1.367(3)
N^1-C^5	1.362(2)	1.392(3)	1.346(2)	1.379(4)	1.371(3)
N^1-CF$_3$	1.406(2)	1.403(3)	1.416(2)	1.419(4)	1.422(4)
Bond angles [°]					
C^5-N^1-N^2	112.5(1)	112.9(2)	112.9(1)	113.0(3)	111.1(2)
C^5-N^1-CF$_3$	128.0(1)	127.9(2)	128.9(1)	129.0(3)	129.5(2)
N^2-N^1-CF$_3$	119.3(1)	118.8(2)	118.1(1)	117.4(3)	119.3(2)
Torsion angles [°]					
C^3/N^3-N^2-N^1-CF$_3$	−176.2(1)	−174.8(2)	177.2(1)	−174.2(3)	−176.6(2)

3.5.4 Ongoing Work

A mild and efficient method for the direct electrophilic *N*-trifluoromethylation of various substituted electron-rich nitrogen heterocycles was developed, which provides ready access to a wide variety of stable N-CF$_3$ compounds. Moderate to good yields were obtained when the well soluble silylated azoles, prepared by an *in situ* silylation sequence, were used in concentrated solutions. Since the substrate must be present in high concentration for an efficient conversion to the desired product, easily soluble substrates are needed and cases where this condition is not fulfilled result in low yields. Although silylation enhances the solubility in some cases drastically, further efforts to optimize the reaction conditions for less soluble substrates are needed, for instance, by working with solvent mixtures. A further drawback that should be addressed is the low or missing conversion of substrates with sterically demanding substituents.

Indoles and hydroxyquinolines are important chemical entities for the pharmaceutical industry. Initial attempts to trifluoromethylate 2-methylindol and quinolone **93** under the standard conditions developed in Section 3.4.1 were carried out and the results are shown in Scheme 22 and Scheme 23, respectively.

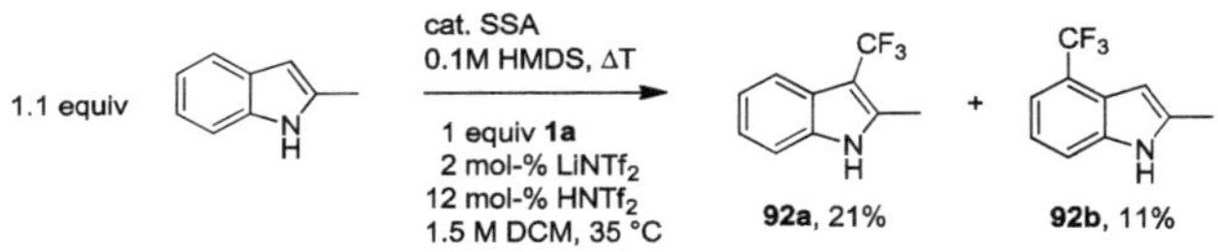

Scheme 22. Trifluoromethylation of 2-methylindole; yields calculated based on integration of ^{19}F NMR signals using $C_6H_5CF_3$ as internal standard.

In previous studies, similar to benzimidazoles, indoles were preferentially trifluoromethylated at the α-position with respect to the nitrogen.[54c, 55a] Therefore, 2-methylindole, blocked in 2-position, was allowed to react under the standard conditions. Disappointingly, the aromatic trifluoromethylated compounds **92a** and **92b** were formed as main products in low yield among other sideproducts. They were separated from the product mixture by flash column chromatography and their structures assigned by multinuclear NMR spectroscopy.

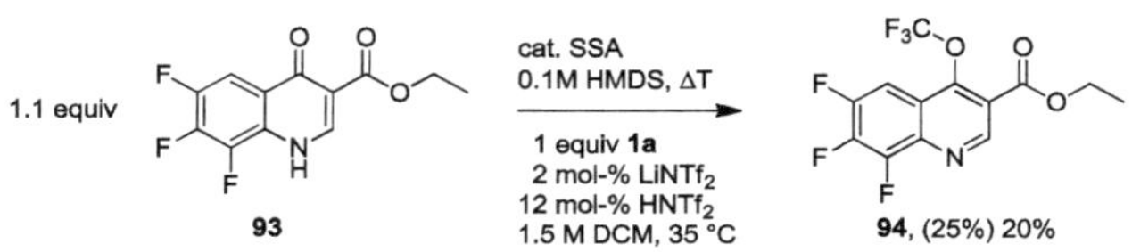

Scheme 23.Trifluoromethylation of quinolone **93**; yield in brackets calculated based on integration of ^{19}F NMR signals using $C_6H_5CF_3$ as internal standard.

Compound **93** is an important intermediate in the synthesis of 8-flouro norfloxacin derivatives that have been proven to exhibit increased potency in their antibacterial activity against *Klebsiella pneumonia* and methicillin or methicillin & vancomycin resistant *Staphylococcus aureus* in comparison to norfloxacin and ciprofloxacin.[114] Therefore, this important pharmaceutical precursor was applied to the standard conditions for the direct electrophilic *N*-trifluoromethylation with the aim of accessing 1-(trifluoromethyl)quinolines. Unfortunately, as shown in Scheme 23, the selectivity of the trifluoromethylation is towards oxygen rather than nitrogen, and **94** is isolated in 24% yield after purification. The structure was assigned by ^{19}F^{1}H HOESY correlation and subsequently by single crystal X-ray analysis. The corresponding spectrum, as well as the ORTEP representation are shown in Figure 20.

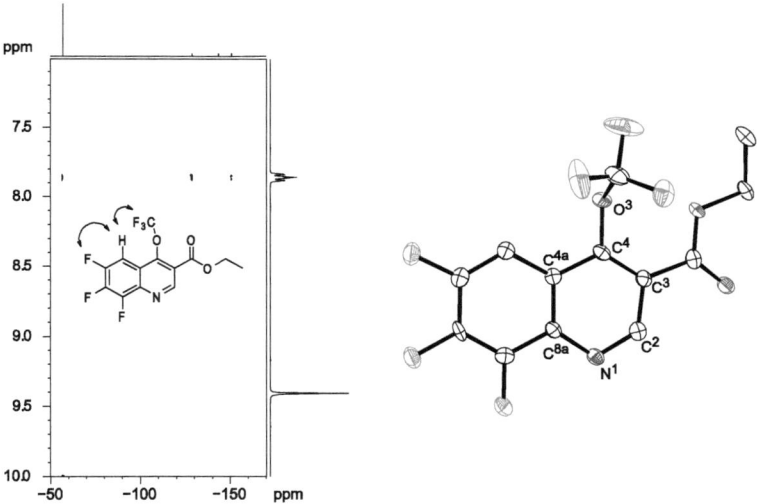

Figure 20. ^{19}F^{1}H HOESY correlation spectrum and ORTEP representation of **94**. Hydrogen atoms are omitted for clarity, thermal ellipsoids set to 50% probability. Selected bond lengths [Å], bond angles [°] and torsion angles [°]: O^3-CF_3 1.358(5), O^3-C^3 1.392(4), N^1-C^2 1.312(5), F_3C-O^3-C^4 118.4(3), C^2-N^1-C^{8a} 116.7(3), F_3C-O^3-C^4-C^3 –91.2(5), F_3C-O^3-C^4-C^{4a} 94.4(4).

In a ^{1}H^{1}H NOESY experiment it was determined that hydroxyquinoline **93** is preferentially silylated at the hydroxyl group, rather than on nitrogen. In addition, the ^{19}F^{1}H HOESY spectrum of silyated quinoline shows only a cross peak between the aromatic proton and fluorine, and no interaction between an aromatic fluorine and the protons of the silyl group is observed. It may be possible to protect the carbonyl group to circumvent this intrinsic problem. However, compound **93** is mainly present in its hydroxy tautomer form.

Inspired by the above results, the conditions were also applied to 2,4,6-trimethylphenol, expecting to improve the direct *O*-trifluoromethylation of phenols, previously examined in our group.[57a] The results shown in Scheme 24 stand in contrast to the earlier reports, where the *para*-methyl group was trifluoromethylated in 49% yield upon reaction with a slight excess of **1b** in DCM at room temperature. Although the reaction presented in Scheme 24 showed no improvement in the direct *O*-trifluoromethylation of the phenol, **95** is formed highly selectively, other trifluoromethylated products being formed in less than 4% each.

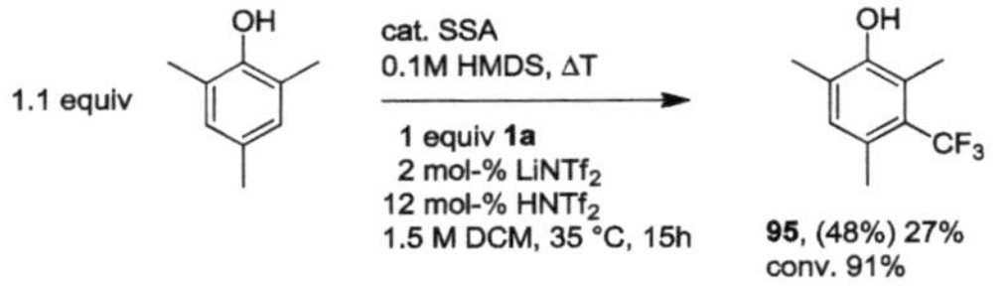

*[Scheme showing trifluoromethylation of 2,4,6-trimethylphenol: 1.1 equiv of 2,4,6-trimethylphenol + 1 equiv **1a**, cat. SSA, 0.1M HMDS, ΔT, 2 mol-% LiNTf$_2$, 12 mol-% HNTf$_2$, 1.5 M DCM, 35 °C, 15h → **95**, (48%) 27% conv. 91%]*

Scheme 24. Trifluoromethylation of 2,4,6-trimethylphenol; yield in brackets calculated based on integration of ^{19}F NMR signals using $C_6H_5CF_3$ as internal standard.

In conclusion it was found that the silyl group does not always direct the outcome of the reaction and that the new conditions involve a new reaction mechanism. Although these preliminary experiments did not yield the desired results, they are still valuable and open the door to new exciting reactions or possible improvements of existing procedures.

3.6 Conclusion and Outlook

In this chapter two methods for the preparation of N-CF$_3$-groups were presented, first the Ritter-type reaction and second the direct *N*-trifluoromethylation of various electron rich heterocycles. These methods provide ready access to a wide variety of stable N-CF$_3$ compounds, as opposed to the preparation of unstable *N*-(diphenylmethylene)-1,1,1-trifluoromethaneamine, described in Section 3.2. All new compounds were fully characterized *inter alia* by 2D-NMR spectroscopy and single crystal X-ray diffration in order to better understand the isomer distribution of the trifluoromethylated products obtained. To gain a deeper insight into the actual processes involved in the reactions, preliminary kinetic experiments as well as computational studies (in collaboration with PD H. P. Lüthi) were carried out, which led to a better understanding of the mechanistic aspects of these new reactions. The products formed in the Ritter-type reaction, *N*-(trifluoromethyl)imidoyl azoles, are rare compounds which are otherwise very difficult to access and their synthetic utility and applications are currently under investigation. First attempts to directly *N*-trifluoromethylate substrates other than azoles did not lead to the desired results; 2-methylindol was preferentially trifluoromethylated at carbon aromatic sites, whereas the reaction with quinoline **93** afforded an *O*-trifluoromethylated product.

At the moment, attempts to circumvent these problems are on-going and finally these results may lead to new reactions or improve existing methods, such as the aromatic trifluoromethylation of heteroarenes or the direct electrophilic trifluoromethylation of phenols.

4 General Conclusion, Comments and Outlook

This thesis presents an X-ray structure correlation and reactivity study on hypervalent iodine compounds designed to guide the search for more effective reagents for the electrophilic trifluromethylation of hard nucleophiles. Therefore, several new five- and a six-membered heterocyclic 1-chloro-λ^3-iodanes, including two cationic species were synthesized, as well as three new trifluoromethylating agents. The structural features were correlated to the relative reactivities towards *para*-toluene sulfonic acid hydrate. Unfortunately, no conditions were found to convert the cationic 1-chloro-λ^3-iodanes into the corresponding trifluoromethyl derivatives. Furthermore, secondary intermolecular contacts in the crystal packing complicated the structure correlation. Therefore, the results of this study should be considered as a qualitative guide. Although a vague trend is observed; it seems that reagents with weakened I-O bonds are more reactive towards these substrates. These conclusions are in agreement with the observation that the I-O bond is elongated in 1-(trifluoromethyl)-λ^3-iodanes upon protonation.[37] Furthermore, this activation allowed to expand the substrate scope of this class of reagents to *O*-centered nucleophiles.[57b, 57c] In this work, the application range was further expanded to *N*-centered nucleophiles considering the same fundamental aspects of reagent activation. Direct *N*-trifluoromethylation is, with one exception,[50] unprecedented. In fact, the construction of an N-CF_3 unit is particularly challenging and *N*-trifluoromethylated compounds are therefore extremely rare and scarcely studied.

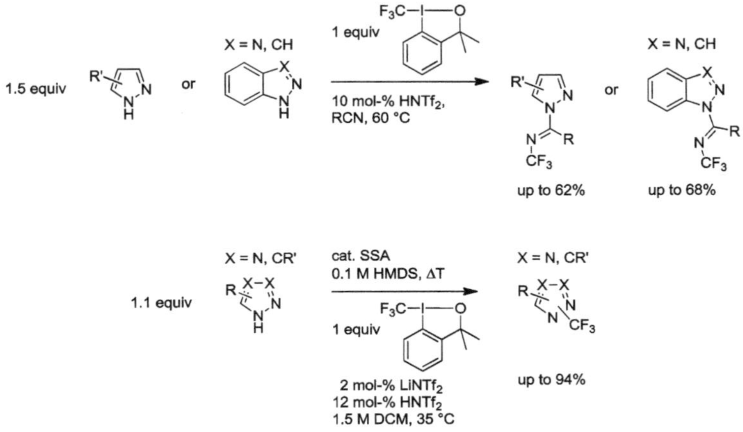

Scheme 25. Ritter-type (upper) and direct (lower) *N*-trifluoromethylation of azoles utilizing hypervalent iodine(III) reagents.

General Conclusion, Comments and Outlook

In this thesis, two acid catalyzed methods for the direct *N*-trifluoromethylation of organonitrogen compounds were presented: the preparation of *N*-(trifluoromethyl)imdoyl azoles by a Ritter-type reaction and direct electrophilic *N*-trifluoromethylation of electron rich heterocycles, as shown in Scheme 25.

In the first method, *N*-trifluoromethylated amidines are formed in a Ritter-type reaction using an azole and reagent **1a** as trifluoromethylating agent in acetonitrile with acid catalysis. The reaction can also be carried out in propionitrile instead of acetonitrile; the corresponding *N*-trifluoromethylated imine is obtained with only slightly lower yield. However, the yield drops significantly when more sterically demanding and conjugated nitriles and/or azoles are used. Furthermore, this reaction is limited to azoles. If the reaction is carried out with benzophenone imine, unstable direct *N*-trifluoromethylated imine was observed instead. Based on preliminarily kinetic experiments as well as computational studies (as part of a collaboration with PD H. P. Lüthi) a reaction mechanism for the Ritter-type reaction is proposed. The second reaction shown in Scheme 25 corresponds to the direct *N*-trifluoromethylation of various electron-rich heterocycles under mild conditions in an efficient manner. Thereby, not only is the reagent activated by the addition of a Brønsted acid, but also the substrate by silylation. *In situ* silylated azoles were *N*-trifluoromethylated in moderate to excellent yields, as long as the reactions were carried out in high concentrations in solvents which are not components of a Ritter-type reaction. As in the case of the latter transformation, the substrate scope is limited to azoles, although a broader range including tetrazoles and benzimidazole has been shown to undergo the transformation. First attempts to directly *N*-trifluoromethylate other substrates (2-methylindole and a 4-hydroxyquinoline derivative) did not lead to the desired results. At the moment, attempts to circumvent these problems are on-going. Nevertheless, we have shown that the substrate scope can be expanded to nitrogen nucleophiles for the electrophilic trifluoromethylation by hypervalent iodine(III) compounds, a transformation that is impossible starting from other bench stable reagents. To test the stability of such *N*-trifluoromethylated compounds remains a task for the future.

To predict the future of this chemistry is a futile exercise. Or to quote Master Yoda from the world famous movie "Star Wars": *the dark side clouds everything; impossible to see the future is.*[115] The topic is just too complex and the variety is enormous. However, some targets for the near future include broadening the substrate scope to simple amines, tertiary alcohols as well as phenols. Furthermore, MacMillan[64] and Kieltsch[99] have shown that the enantioselective trifluoromethylation utilizing these reagents is possible. Some further efforts in this direction would be desirable. Moreover, we are currently developing new reagents based on hypervalent iodine(III) to broaden the scope of the moiety transferred to CF_2H to allow access to the equally important class of CF_2H substituted compounds.

5 Experimental Part

5.1 General Remarks

Part of the procedures described in this chapter including spectroscopic data have been reported as supporting information of the published papers.[116]

Crystallographic tables are given in the appendix. Furthermore, with the exception of 1-(dichloro-3-iodanyl)-2-(1-fluoro-1-methylethyl)benzene (**25a**) and ethyl 6,7,8-trifluoro-4-(trifluoromethoxy)quinoline-3-carboxylate (**94**) the structural data can be obtained from *The Cambridge Crystallographic Data Centre* via www.ccdc.ac.uk/data_request/cif.

1-(2-Iodophenyl)cyclohexanol (**5**), 1-chlorospiro[1λ^3,2-benziodaoxole-3.1'-cyclohexane] (**11**), 1-(trifluoromethyl)spiro[1λ^3,2-benziodaoxole-3.1'-cyclohexane] (**27**), 9-(2-iodophenyl)bicycle[3.3.1]nonan-9-ol (**6**) and 1-chlorospiro[1λ^3,2-benziodaoxole-3.9'-bicyclo[3.3.1]nonane] (**12**) were prepared and characterized by *Dr. Ján Cvengroš*, 2-(2-iodphenyl)-1-methoxy-2-propanol (**7**) and 1-chloro-3-methoxymethyl-3-methyl-1H,3H-3-dihydro-1,2-benziodoxol (**13**) by *Philip Battaglia*, 1-trifluoromethyl-3-methyl-3-phenyl-1H,3H-3-dihydro-1,2-benziodoxol (**28**) and 1-chloro-3-isopropyl-3-phenyl-1,3-dihydro-1,2-benziodoxol (**15**) by *Dr. Raffael Koller* and 8-chloro-8λ^3-ioda-7λ^5-azatricyclo-[7.4.0.0^{2,7}]trideca-1(9),2,4,6,10,12-hexaen-7-ylium tetrafluoro-λ^4-borane (**21**) by *Nico Santschi*.

The direct *N*-trifluoromethylation of triazoles was carried out in collaboration with *Remo Senn* and *Barbara Czarniecki*.

5.1.1 Techniques

Reactions were carried out under an argon atmosphere using standard Schlenk techniques and in a glovebox with an N_2 atmosphere. Unless explicitly indicated, the solvents were freshly distilled from an appropriate drying agent: THF, Et_2O, hexane from Na/benzophenone; pentane from Na/benzophenenone/diglyme; MeOH, EtOH, DCM, MeCN from CaH_2; toluene from Na; EtCN from P_2O_5. DCM-d_2 and MeCN-d_3 were bulb-to-bulb distilled from CaH_2 and degassed by three freeze-pump-thaw cycles.

Neutral and basic aluminium oxide activity I was purchased from *ICN Biomedicals GmbH*, silica gel 60 (230-400 mesh) from *Fluka* and Florisil (100-200 mesh) from *ABCR*. TLC-plates were obtained from *Merck* (silica gel 60 F_{254}).

5.1.2 Analytical Methods

^{1}H, ^{13}C, ^{15}N and ^{19}F **NMR spectra** were recorded on *Bruker DPX* 250, *DPX* 300, *DPX* 400, *DPX* 500 and *Avance* 700 spectrometers. The samples were measured as

Experimental Part

solutions in the given solvent at room temperature in non-spinning mode. ^{1}H and ^{13}C chemical shifts are referenced relative to external tetramethylsilane. ^{19}F NMR spectra were referenced to external CFCl$_3$. ^{15}N NMR spectra were referenced to external NH$_3$. Chemical shifts for ^{15}N were determined by measurement of ^{1}H^{15}N HMQC and/or ^{19}F^{15}N HMQC spectra. The multiplicities of the signals are abbreviated as follows: s = singlet, d = doublet, t = triplet, q = quartet, hept = heptet, ψ = pseudo/appears as, br = broad, m = multiplet, u = unresolved. For the assignment of the ^{1}H and ^{13}C chemical shifts standard ^{1}H^{13}C HMQC, ^{1}H^{13}C HMBC and ^{1}H^{1}H COSY experiments were measured. For the assignment of regioisomers ^{1}H^{1}H NOESY and ^{19}F^{1}H HOESY spectra were recorded. The ^{1}H^{15}N HMQC spectra were acquired on a *Bruker DPX* 400 or *Avance* 700 spectrometer equipped with a multinuclear inverse probe. A relaxation delay of 800 ms was applied and a defocusing delay of 100 ms was chosen, corresponding to a coupling constant of 5 Hz. The number of scans per increment was 16 (2k data points), and 256 experiments were acquired in the second dimension. Total experimental time was ca. 1.5 h. The ^{19}F^{15}N HMQC spectra were acquired on *Bruker DPX* 400 or *Avance* 700 spectrometer equipped with a multinuclear inverse probe. A relaxation delay of 800 ms was applied and a defocusing delay of 40 ms was chosen, corresponding to a coupling constant of 12.5 Hz. The number of scans per increment was 16 or 32 (2k data points), and 256 or 512 experiments were acquired in the second dimension. Total experimental time was 1.5 to 4.5 h. ^{1}H^{1}H NOESY spectra were acquired on *Bruker DPX* 400, *DPX* 500 and *Avance* 700 spectrometers equipped with a multinuclear inverse probe. A relaxation delay of 800 ms was applied and the mixing time was 600 or 800 ms. The number of scans per increment was 16 or 32 (2k data points) and 512 experiments were acquired in the second dimension. Total experimental times were between 6-10 h. The ^{19}F^{1}H HOESY spectra were acquired using the standard four-pulse sequence on a *Bruker DPX* 400 spectrometer equipped with a doubly tuned (^{1}H, ^{19}F) TXI probe. A relaxation delay of 1 s was applied and the mixing time was 800 ms. Typically, 16 to 32 transients were acquired into 2k data points for each of the 256, 512 or 1k increments in t$_1$. Total experimental times were between 5 and 12 h. **Infrared spectra** were recorded on a *Thermo Fisher Scientific Nicolet 6700 FT-IR Pike Technologies GladiATR*^{TM}. **Melting points** were measured on a *Büchi Melting Point B-540* apparatus or were determined by DSC (10 °C/min) onset on a *Mettler Toledo Polymer DSC* apparatus using the *Mettler Toledo STAR^{e}* program. **Boiling points** were determined by DSC (10 °/min) onset on a *Mettler Toledo Polymer DSC* apparatus using the *Mettler Toledo STAR^{e}* program. Temperatures are given in degree Celsius (°C) and are uncorrected. **Mass spectra** were measured by the MS service of the *Laboratorium für Organische Chemie (ETH Zürich)*. **GC-MS** measurements were performed on a *thermo Finnigan Trace GC 2000/Trace MS* equipped with a *Phenomenex Zebron* ZB-column (lenghth: 30

m, 0.25 mm inner diameter, 0.25 µm coating thickness) coupled to a quadrupole mass filter. Helium was used as the carrier gas with a constant flow of 1.2 mL/min. separation of the injected species was achieved using the denoted temperature program and retention times t_R are given in minutes (min). **Elemental analysis** was carried out by the *Laboratory of Microelemental Analysis* of the *ETH Zürich*. Intensity data for **single crystals** glued to a glass capillary were collected at 100 K or 200 K on a *Bruker* SMART APEX platform with CCD detector and graphite monochromated Mo-K$_\alpha$-radiation (λ = 0.71073 Å). The program SMART was used for data collection and integration was performed with the software SAINT+.[117] The structures were solved by direct methods using the program SHELXS-97,[118] subsequent refinement and all further calculations were carried out using SHELXL-97.[119] All non-hydrogen atoms were refined anisotropically using weighted full-matrix least-squares on F^2. The hydrogen atoms were included in calculated positions and treated as riding atoms using SHELXL default parameters. Absorption correction was applied (SADABS)[120] and weights were optimized in the final refinement cycles. The standard uncertainties (s.u.) are rounded according to the "Note for Authors" of *Acta Crystallographica*.[121]

5.1.3 Chemicals

Commercial compounds were obtained from *ABCR*, *Acros*, *Alfa Aesar*, *Fluka*, *Lancaster*, *Sigma-Aldrich*, *Strem* and *TCI* and used as received without any further purification unless stated otherwise.

Ethyl 1,2,3,4-tetrahydro-1,4-methanonaphthalene-1-carboxylate (**18**),[122] 2-(2-iodo-phenyl)-4,4-dimethyl-4,5-dihydro-1,3-oxazole (**22**),[116a] 2-(2-iodophenyl)pyridine (**23**),[116a] 2-(2-iodophenyl)propan-2-ol (**4a**),[123] 1-chloro-1,3-dihydro-3,3-dimethyl-1,2-benziodoxole (**2a**),[123] 1-chloromethyl-3-methyl-3-phenyl-1H,3H-3-dihydro-1,2-benziodoxol (**14**),[72] 1-trifluoromethyl-1,3-dihydro-3,3-dimethyl-1,2-benziodoxol (**1a**),[123] chlorofluoro(trifluoromethyl)-(9CI)-iodine (CF$_3$I(Cl)F),[83] 1,1,3,3-tetrakis(trifluoromethane-sulfonyl)propane ((Tf$_2$CH)$_2$CH$_2$),[124] silica sulfuric acid (SSA),[111] 3-(1-adamantyl)-1H-pyrazole (**43**),[125] 3-(2,4,6-trimethylphenyl)-1H-pyrazole (**45**),[126] ethyl 3-methyl-1H-pyrazole-4-carboxylate (**67**),[127] 4-benzyl-1H-pyrazole ((**68**), recrystallized from hot heptanes),[128] 3,5-dimethyl-1-(trimethylsilyl)-1H-pyrazole (**69**),[129] 5-methyl-3-(2,4,6-trimethylphenyl)-1H-pyrazole (**71**),[130] were synthesized as reported in the literature.

1-Chloro-1,3-dihydro-3,3-bis(trifluoromethyl)-1,2-benziodoxole (**1c**)[131] was kindly provided by *Dr. Jan M. Welch*.

5.2 Hypervalent Iodine Compounds

1-(2-Iodophenyl)cyclohexanol (5)

Anhydrous $CeCl_3$ (713 mg, 2.89 mmol, 1.5 equiv) was suspended in dry THF (8 mL) and stirred for 12 h at room temperature. In a second Schlenk flask, 1,2-diiodobenzene (955 mg, 2.89 mmol, 1.5 equiv) was dissolved in dry THF (10 mL) under argon. After cooling to −30 °C, iPrMgCl (1.45 mL, 2.0 M solution in THF, 2.89 mmol, 1.5 equiv) was added dropwise and the resulting orange colored mixture was warmed to −20 °C over a period of 20 minutes. The reaction was monitored by GC-MS. Both flasks were cooled to −78 °C and the freshly prepared Grignard reagent was added slowly to the $CeCl_3$ suspension by means of a syringe. The mixture was warmed to room temperature to ensure the complete formation of organocerium species by transmetallation. After cooling back to −78 °C, cyclohexanone (0.2 mL, 1.92 mmol) was added and the mixture was allowed to warm to room temperature overnight. After dilution with Et_2O (20 mL), the reaction mixture was treated with saturated aqueous NH_4Cl (20 mL) while being cooled in an ice bath. The aqueous phase was extracted twice with Et_2O (20 mL). The combined organic layers were dried over anhydrous Na_2SO_4 and the solvent evaporated in vacuo. After purification of the crude product by flash chromatography (SiO_2, hexane:EtOAc 50:1 then 5:1), **5** (384 mg, 66%) was isolated as a white powder. **^{1}H NMR** (300 MHz, $CDCl_3$): δ = 7.99 (t, $J_{H,H}$ = 7.9 Hz, 1H, $C_{Ar}H$), 7.62 (t, $J_{H,H}$ = 7.9 Hz, 1H, $C_{Ar}H$), 7.35 (t, $J_{H,H}$ = 7.8 Hz, 1H, $C_{Ar}H$), 6.92 (t, $J_{H,H}$ = 7.4 Hz, 1H, $C_{Ar}H$), 2.32 (s, 1H, OH), 2.02-2.21 (m, 4H, CH), 1.69-1.91 (m, 5H, CH), 1.28-1.42 (m, 1H, CH); **^{13}C{^{1}H} NMR** (75 MHz, CDCl3): δ = 148.4, 143.0, 128.6, 128.2, 127.0, 93.5 (CI), 74.0 (COH), 36.0, 25.3, 22.0; **HRMS (EI)**: calcd m/z for $C_{12}H_{15}IO$: 302.0168 [M$^+$], found 302.0162 [M$^+$]; **CAS**: 1193603-08-5.

1-Chlorospiro[1λ^3,2-benziodaoxole-3.1'-cyclohexane] (11)

A round bottomed flask was charged with alcohol **5** (380 mg, 1.26 mmol), and CH_2Cl_2 (4 mL) and cooled to 0 °C. To the slightly yellow reaction mixture, tBuOCl (148 μL, 1.27 mmol, 1.01 equiv) was added and the mixture was warmed to room temperature overnight. The solution was concentrated and the crude product recrystallized from CH_2Cl_2 to give a yellow solid (403 mg, 95%). Single crystals for X-ray analysis were obtained by diffusion of pentane into a saturated CH_2Cl_2 solution. **^{1}H NMR** (300 MHz, $CDCl_3$): δ = 8.04 (d, $J_{H,H}$ =7.5 Hz, 1H, $C_{Ar}H$), 7.50-7.59 (m, 2H, $C_{Ar}H$), 7.18 (d, $J_{H,H}$ = 6.9 Hz, 1H, $C_{Ar}H$), 1.90-1.94 (m, 2H, CH), 1.58-1.82 (m, 7H, CH), 1.22-1.36 (m, 1H, CH); **^{13}C{^{1}H} NMR** (75 MHz, $CDCl_3$): δ = 149.0, 130.8, 130.5, 128.5, 126.1, 115.2 (CI), 86.4 (CO), 37.2, 25.4, 22.1;

HRMS (EI): calcd m/z for $C_{12}H_{14}ClIO$: 335.9778 [M$^+$], found 335.9771 [M$^+$]; **CAS:** 1240913-16-9; **CCDC:** 771239.

1-(Trifluoromethyl)spiro[1λ^3,2-benziodaoxole-3.1'-cyclohexane] (27)

Chlorobenziodaoxole **11** (1.2 g, 3.5 mmol) was dissolved in CH$_3$CN (25 mL) under argon and AgOAc (0.44 g, 3.7 mmol, 1.05 equiv) was added. The resulting suspension was stirred for 3 h, filtered and concentrated to yield the corresponding acetate which was used without further purification. The acetate was dissolved in CH$_3$CN (30 mL) and TMSCF$_3$ (0.8 mL, 5.3 mmol, 1.5 equiv) was added, followed by a solution of TBAT (3.8 mg, 7 μmol, 0.2 mol-%) in CH$_3$CN (2 mL) at −17 °C. The resulting mixture was stirred at that temperature for 20 h. The solution was allowed to warm to −12 °C, additonal TMSCF$_3$ (0.13 mL, 0.88 mmol, 0.25 equiv) was added and the stirring was then continued for 24 h at ambient temperature. Pentane (20 mL) was added and the solution was filtered through cotton and concentrated. The crude product was purified by flash-chromatography (Alox, hexane:EtOAc = 50:1) to give **27** as a white solid (0.48 g, 37%). X-ray quality crystals were obtained by sublimation under high vacuum (0.015 mbar, 60 °C). **^{1}H NMR** (300 MHz, CDCl$_3$): δ = 7.52-7.56 (m, 2H, C$_{Ar}$H), 7.40-7.45 (m, 2H, C$_{Ar}$H), 1.90-1.94 (m, 2H, CH), 1.63-1.78 (m, 7H, CH), 1.25-1.29 (m, 1H, CH); **^{13}C{^{1}H} NMR** (75 MHz, CDCl$_3$): δ = 149.5, 130.5, 129.5, 127.9 (q, $J_{C,F}$ = 2.7 Hz), 127.3, 111.3 (q, $J_{C,F}$ = 3 Hz, CI), 110.9 (q, $^1J_{C,F}$ = 397.0 Hz, CF$_3$), 78.1 (CO), 38.5, 25.7, 22.4; **^{19}F NMR** (188 MHz, CDCl$_3$): δ = −40.3; **HRMS (EI):** calcd m/z for $C_{13}H_{14}F_3IO$: 370.0041 [M$^+$], found: 370.0044 [M$^+$]; **CAS:** 1240913-38-5; **CCDC:** 771240.

9-(2-Iodophenyl)bicyclo[3.3.1]nonan-9-ol (6)

Starting from bicyclo[3.3.1]nonan-9-one (183 mg, 1.3 mmol), **6** was prepared in analogy to compound **5**. After purification by column chromatography (hexane:EtOAc 50:1 then 10:1) the title compound was obtained as a white powder (363 mg, 82%). **^{1}H NMR** (300 MHz, CDCl$_3$): δ = 8.03 (t, $J_{H,H}$ = 7.7 Hz, 1H, C$_{Ar}$H), 7.62 (t, $J_{H,H}$ = 7.5 Hz, 1H, C$_{Ar}$H), 7.35 (t, $J_{H,H}$ = 7.5 Hz, 1H, C$_{Ar}$H), 6.91 (t, $J_{H,H}$ = 6.9 Hz, 1H, C$_{Ar}$H), 3.03 (br s, 2H, CH), 2.40-2.51 (m, 3H, CH), 1.57-2.03 (m, 10H, CH), 1.41 (m, 1H, CH); **^{13}C{^{1}H} NMR** (75 MHz, CDCl$_3$): δ = 145.4, 143.9, 129.4, 128.6, 127.6, 93.6 (CI), 76.1 (COH), 34.8, 29.9, 27.6, 20.8, 19.9; **HRMS (EI):** calcd m/z for $C_{15}H_{19}IO$: 342.0481 [M$^+$], found: 342.0477 [M$^+$]; **CAS:** 1240913-10-3.

Experimental Part

1-Chlorospiro[1λ^3,2-benziodaoxole-3.9'-biyclo[3.3.1]nonane] (12)

Starting from *ortho*-iodobenzyl alcohol **6** (356 mg, 1.04 mmol), **12** was prepared in analogy to **11** yielding a yellow solid (360 mg, 92%). Single crystals for X-ray analysis were obtained by diffusion of pentane into a saturated CH$_2$Cl$_2$ solution. **^{1}H NMR** (300 MHz, CDCl$_3$): δ = 8.16 (d, $J_{H,H}$ = 6.0 Hz, 1H, C$_{Ar}$H), 7.81 (d, $J_{H,H}$ = 6.0 Hz, 1H, C$_{Ar}$H), 7.51 (m, 2H, C$_{Ar}$H), 1.52-2.35 (m, 14H, CH); **^{13}C{^{1}H} NMR** (75 MHz, CDCl$_3$): δ = 147.7, 129.7, 129.5, 129.0, 128.9, 118.9 (C*I*), 88.6 (*C*O), 36.5, 28.8, 28.7, 20.1, 20.0; **HRMS (EI)**: calcd *m/z* for C$_{15}$H$_{18}$ClIO: 376.0091 [M$^+$], found: 376.0086 [M$^+$]; **CAS**: 1240913-19-2; **CCDC**: 771241.

2-(2-Iodphenyl)-1-methoxy-2-propanol (7)

1,2-Diiodobenzene (0.2 mL, 1.5 mmol) was dissolved in THF (5 mL) and cooled to –30° C. After the addition of *i*PrMgCl (1 mL, 2 M in THF, 2 mmol, 1.3 equiv), the solution was allowed to warm to –20 °C over 15 min and stirred at this temperature for an additional 20 min. At –78 °C, methoxyacetone (0.14 mL, 1.5 mmol, 1 equiv) in THF (2 mL) was added and the suspension was stirred at –78 °C for 4 h and 3 h at room temperature The resulting yellow solution was diluted with Et$_2$O (10 mL), and saturated aqueous NH$_4$Cl was added while the mixture was cooled on an ice bath. The aqueous phase was extracted with Et$_2$O (3x 10 mL) and the combined organic phases were washed with H$_2$O (20 mL) and brine (20 mL). The organic phase was then dried over MgSO$_4$, filtered and the solvent removed under reduced pressure. After purification by flash chromatography (SiO$_2$, *c*-hexane/EtOAc 6:1), compound **7** (0.105 g, 24%) was isolated as a colourless oil. R_f (c-hexane/EtOA 6:1): 0.24; **^{1}H NMR** (300 MHz, CDCl$_3$): δ = 7.98(d, $J_{H,H}$ =7.8 Hz, 1H, C$_{Ar}$H), 7.77 (dd, $J_{H,H}$ =7.8 Hz, $J_{H,H}$ = 1.2 Hz, 1H, C$_{Ar}$H), 7.37(t, $J_{H,H}$ = 8.1 Hz, 1H, C$_{Ar}$H), 6.92(dt, $J_{H,H}$ = 8.1 Hz, $J_{H,H}$ = 1.2 Hz, 1H, C$_{Ar}$H), 3.95(dd, $J_{H,H}$ = 16.5 Hz, 9.5 Hz, 2H, CH$_2$), 3.39(s, 3H, OCH$_3$), 3.29 (s, 1H, OH), 1.71(s, 3H, CH$_3$), **^{13}C{^{1}H} NMR** (63MHz; CDCl$_3$): δ = 162.3, 146.2, 142.7, 128.8, 128.11, 128.06, 93.0, 77.5, 74.6, 59.3, 24.7; **HRMS (EI)**: calcd *m/z* for C$_{10}$H$_{13}$IO$_2$: 291.9955 [M$^+$], found: 291.9953 [M$^+$]; **CAS**: 1240913-12-5.

1-Chloro-3-methoxymethyl-3-methyl-1*H*,3*H*-λ^3-dihydro-1,2-benziodoxol (13)

Starting from *ortho*-iodobenzyl alcohol **7** (105 mg, 0.34 mmol) **13** was prepared in analogy to **11** yielding a yellow solid (63.4 mg, 58%). **^{1}H NMR** (300 MHz, CDCl$_3$): δ = 8.03 (d, $J_{H,H}$ = 9.0 Hz, 1H, C$_{Ar}$H), 7.57 (t, $J_{H,H}$ = 9.3 Hz, 2H, C$_{Ar}$H), 7.22 (d, $J_{H,H}$ = 7.5 Hz, 1H, C$_{Ar}$H), 3.55 (dd,

$J_{H,H}$ =25.7 Hz, $J_{H,H}$ = 11.4 Hz, 2H, CH_2), 3.35 (s, 3H, OCH_3), 1.55 (s, 3H, CH_3); $^{13}C\{^1H\}$ **NMR** (63 MHz, $CDCl_3$): δ = 146.4, 130.8, 130.7, 128.3, 127.1, 115.2, 86.1, 79.7, 59.8, 25.2; **HRMS (EI)** calcd m/z for C_8H_7ClIO: 280.9230 [M^+-C_2H_5O], found: 280.9225246 [M^+-C_2H_5O]; **CAS**: 1240913-21-6; **CCDC**: 771244.

1-Trifluoromethyl-3-methyl-3-phenyl-1H,3H-λ^3-dihydro-1,2-benziodoxol (28)

A 100 mL Schlenk flask was charged with KOAc (1.43 g, 14.5 mmol, 1.65 equiv) which was dried under vacuum using a heat gun. Under counter flow of argon, chloride **8** (3.16 g, 8.81 mmol, 1.0 equiv) was added. CH_3CN (25 mL) was added via syringe to give a yellow suspension. After stirring for 1 h at room temperature the resulting white suspension was filtered into a 100 mL Schlenk flask under argon. To the colorless solution, further CH_3CN (20 mL) was added. After cooling to –19 °C, $TMSCF_3$ (2.1 mL, 14 mmol, 1.6 equiv) was added, followed by TBAT (5.4 mg, 0.088 mmol, 1 mol-%) in CH_3CN (2 mL). The reaction mixture was stirred for 24 h at –16 °C, then warmed to –12 °C, at which temperature further $TMSCF_3$ (0.33 mL, 2.2 mmol, 0.25 equiv) was added. The reaction mixture was warmed to room temperature and the solvent was removed under vacuum. Dry pentane (40 mL) was added to the remaining brown solid, and the resulting mixture was filtered through a pad of dry Alox in a Young-filter. The clean, colorless solution was partially evaporated until a white solid precipitated. The suspension was cooled to –40 °C and the solvent was decanted. The white crystalline residue was dried under vacuum to yield **28** (2.80 g, 81%) as a white solid.

Enantiomerically pure **28** did not crystallize upon concentration and cooling, therefore the solution was concentrated in vacuuo to give a colorless oil having identical spectroscopic properties to the racemate. ^{1}**H NMR** (300 MHz, $CDCl_3$): δ = 7.63-7.52 (m, 2H, $C_{Ar}H$), 7.49-7.39 (m, 4H, $C_{Ar}H$), 7.36-7.21 (m, 3H, $C_{Ar}H$), 1.86 (s, 3H, CH_3); $^{13}C\{^1H\}$ **NMR** (75 MHz, $CDCl_3$): 147.95, 147.51, 130.49, 130.08 (q, $J_{C,F}$ =0.6 Hz), 129.36, 128.33, 128.11 (q, $J_{C,F}$ =2.8 Hz), 127.10, 126.13, 112.08 (q, $J_{C,F}$ =2.9 Hz), 110.65 (q, $^1J_{C,F}$ =396 Hz, CF_3), 80.19, 30.83. ^{19}**F NMR** (282 MHz, $CDCl_3$): –39.74 (d, J = 1.4 Hz, CF_3); **HRMS(MALDI)** calcd m/z for $C_{15}H_{12}F_3IO$: 92.99573 [MH^+], found 392.9958 [MH^+]; **Elemental Analysis** calcd (%) for $C_{15}H_{12}OF_3I$: C 45.94, H 3.08, F 14.53; found: C 46.00, H 3.10; F 14.49; **CAS**: 1240913-41-0; **CCDC**: 771246.

1-Chloro-3-isopropyl-3- phenyl-1,3-dihydro-1,2-benziodoxol (15)

A round bottom flask was charged with Mg turnings (675 mg, 27.8 mmol, 1.5 equiv) and Et_2O (10 mL). To this suspension was added 2-iodopropane (1.95 mL, 19.4 mmol, 1.05 equiv) at a rate to keep the suspension at reflux. After complete addition of 2-iodopropane the suspension was

Experimental Part

cooled to room temperature. In a second round bottom flask, 2-iodobenzophenone (5.70 g, 18.5 mmol, 1.0 equiv) was dissolved in Et$_2$O (20 mL) and cooled to 0 °C. To this solution, the Grignard reagent was added dropwise to give a dark orange suspension which was stirred at 0°C for a further two hours. The suspension was warmed to room temperature and quenched with a saturated solution of aqueous NH$_4$Cl. The organic and aqueous phases were separated and the aqueous phase was extracted with Et$_2$O. The combined organic phases were dried over MgSO$_4$ and concentrated to give a yellow oil (5.08 g). Purification proved to be cumbersome; therefore the crude reaction mixture was used for the next step without further purification. A round bottom flask was charged with the crude reaction mixture (500 mg, 1.42 mmol, 1.0 equiv) which was dissolved in CH$_2$Cl$_2$ (10 mL) and cooled to 0 °C. To the slight yellow reaction mixture tBuOCl (161 µL, 1.42, 1.0 equiv) was added and the mixture was allowed to warm to room temperature overnight. The solution was concentrated and redissolved in the minimum amount CH$_2$Cl$_2$ and then layered with Et$_2$O resulting, on standing, in bright yellow crystals. The procedure was repeated 2 times and the combined crops of crystals were recrystallized from CH$_2$Cl$_2$ and Et$_2$O to give bright yellow crystals (246 mg. 44% over 2 steps). ^{1}**H NMR** (300 MHz, CDCl$_3$): δ = 8.01 (dd, $J_{H,H}$ = 1.3 Hz, $J_{H,H}$ = 8.0 Hz, 1H, C$_{Ar}$H), 7.64-7.40 (m, 5H, C$_{Ar}$H), 7.39-7.21 (m, 3H, C$_{Ar}$H), 2.70 (hept, $J_{H,H}$ = 6.7 Hz, 1H, CH(CH$_3$)$_2$), 0.93 (d, $J_{H,H}$ = 6.7 Hz, 3H, CH(CH$_3$)$_2$), 0.84 (d, $J_{H,H}$ = 6.8 Hz, 3H, CH(CH$_3$)$_2$); ^{13}**C{^{1}H} NMR** (75 MHz, CDCl$_3$): δ = 147.11, 143.19, 130.50, 130.38, 128.53, 128.27, 128.07, 127.19, 125.64, 116.19, 92.44, 38.84, 17.49, 16.64. **Elemental Analysis** calcd (%) for C$_{16}$H$_{16}$OCl: C 49.70, H 4.17, O 4.14, Cl 9.17; found: C 49.73, H 4.22, O 4.27, Cl 9.25; **CAS**: 1240913-13-6**CCDC**: 771247.

2-(3,4-Dihydro-2H-1,4-methy nonaphthalen-1-yl)-propan-2-ol (19)

In a two-neck round bottom flask (50 mL) equipped with a reflux condenser, CH$_3$MgI (3.8 mL, 3 M in Et$_2$O, 11 mmol, 2.2 equiv) was diluted with Et$_2$O (20 mL). At 0 °C, the ethyl ester **18** (1.133 g, 5.2 mmol) dissolved in Et$_2$O (6 mL) was slowly added. The mixture was refluxed for 2 h and then cooled to 0 °C and a saturated solution of aqueous NH$_4$Cl was slowly added. The mixture was extracted with Et$_2$O (3x 15 mL), the organic phase dried over K$_2$CO$_3$, filtered and the solvent was removed under reduced pressure. After purification by flash chromatography (SiO$_2$, hexane/EtOAc 5:1), compound **19** (0.982 g, 93%) was isolated as a colourless oil. R_f (hexane/EtOA 5:1): 0.4; ^{1}**H NMR** (300 MHz, CDCl$_3$): δ = 7.50-7.47 (m, 1H, C$_{Ar}$H), 7.24-7.10 (m, 3H, C$_{Ar}$H), 3.36 (br s, 1H CH), 2.31-2.01 (m, 2H, CHH$_{eq}$CHH$_{eq}$), 1.74 (dq, $J_{H,H}$ = 8.6 Hz, $J_{H,H}$ = 2.1 Hz, 1H, CHH$_{endo}$), 1.71 (s, 1H, OH), 1.62 (dd, $J_{H,H}$ = 6.9 Hz, $J_{H,H}$ =1.5 Hz, 1H CH$_{exo}$H), 1.54 (s, 3H, CH$_3$), 1.52 (s, 3H, CH$_3$),

1.39-1.17 (m, 2H, CH_{ax}HCH_{ax}H); ^{13}C{^{1}H} NMR (62.9 MHz, CDCl$_3$): δ = 149.4, 146.8, 125.5, 125.3, 121.6, 120.8, 72.3, 62.8, 49.5, 42.9, 29.4, 27.7, 27.6, 27.5; **HRMS (EI)** calcd m/z for C$_{14}$H$_{18}$O: 202.1353 [M$^+$], found 202.1355 [M$^+$]; **Elemental Analysis** calcd (%) for C$_{14}$H$_{18}$O: C 83.12, H 8.97, O 7.91;. found: C 82.94, H 8.98, O 8.15; **CAS**: 1240913-23-8.

2-(8-Iodo-3,4-dihydro-2H-1,4-methanonaphthalen-1-yl)-propan-2-ol (10)

In a two-neck round bottom flask with a reflux condenser, alcohol **19** (1.36 g, 6.7 mmol) was dissolved in n-hexane (27 mL) and TMEDA (2.2 mL, 14.6 mmol, 2.2 equiv) was added. The solution was cooled to –78 °C and sBuLi (11.4 mL, 1.3 M in hexane, 14.8 mmol, 2.2 equiv) was slowly added. The reaction mixture was heated to reflux for 20 h, afterwards it was cooled to –78 °C and 1,2-diiodoethane (0.811 g, 2.9 mmol, 1.3 equiv) in THF (3 mL) was slowly added. The mixture was allowed to warm slowly to room temperature over night (16 h). Saturated aqueous Na$_2$S$_2$O$_3$ was added and the mixture was extracted with Et$_2$O. The organic phase was washed with brine, dried over Na$_2$SO$_4$, filtered and the solvent was removed under vacuum. The residue was purified by flash chromatography (SiO$_2$, hexane/EtOAc 10:1) to give compound **10** (1.18 g, 54%) as a colourless oil. R_f (hexane/EtOAc 10:1): 0.27; ^{1}**H NMR** (300 MHz, CDCl$_3$): δ = 7.70 (dd, $J_{H,H}$ = 7.9 Hz, $J_{H,H}$ = 1.1 Hz, 1H, C$_{Ar}$H), 7.17 (dd, $J_{H,H}$ = 7.0 Hz, $J_{H,H}$ = 1.0 Hz, 1H, C$_{Ar}$H), 6.77 (dd, $J_{H,H}$ = 7.9 Hz, $J_{H,H}$ = 7.1 Hz, 1H, C$_{Ar}$H), 3.38 (br s, 1H, OH), 3.32-3.30 (s, 1H, CH), 2.16-1.98 (m, 2H, CHH$_{eq}$CHH$_{eq}$), 1.71 (s, 3H, CH$_3$), 1.67-1.58 (m, 2H, CH$_2$), 1.38 (s, 3H, CH$_3$), 1.34-1.21 (m, 2H, CH_{ax}HCH_{ax}H); ^{13}C{^{1}H} **NMR** (62.9 MHz, CDCl$_3$): δ = 154.0, 150.6, 139.2, 127.5, 120.8, 89.3, 71.7, 66.5, 50.7, 43.1, 29.7, 29.5, 28.6, 26.8; **HRMS (EI)** calcd m/z for C$_{14}$H$_{17}$IO: 328.0319 [M$^+$], found 328.0317 [M$^+$]; **Elemental Analysis** calcd (%) for C$_{14}$H$_{17}$OI: C 51.24, H 5.22, O 4.88, I 38.67; found: C 51.43, H 5.26, O 4.90, I 38.78; **CAS**: 1240913-24-9.

1-Chloro-3,3-dimethyl-3a,6-methano-3a,4,5,6,-tetrahydro-1H,3H-λ^3-ioda-2-oxaphenalene (16)

Starting from ortho-iodobenzyl alcohol **10** (1.12 g, 3.4 mmol) **16** was prepared in analogy to **11** to yield colourless crystals (1.09 g, 88%). Single crystals for X-ray analysis were obtained by diffusion of pentane into a saturated CH$_2$Cl$_2$ solution. ^{1}**H NMR** (250 MHz, CD$_2$Cl$_2$): δ = 8.25-8.18 (m, 1H, C$_{Ar}$H), 7.37-7.30 (m, 2H, C$_{Ar}$H), 3.41 (u, 1H, CH), 2.14-2.01 (m, 2H, CHH$_{eq}$CHH$_{eq}$), 1.61 (s, 2H, CH$_2$), 1.55 (s, 3H, CH$_3$), 1.44 (s, 3H, CH$_3$), 1.26 (d, $J_{H,H}$ = 7.5 Hz, 2H,

Experimental Part

$CH_{ax}HCH_{ax}H$); $^{13}C\{^1H\}$ **NMR** (63 MHz, CD_2Cl_2): δ = 153.59, 144.73, 129.48, 125.28, 122.55, 106.66, 75.19, 57.20, 49.22, 42.12, 28.78, 28.50, 26.48, 24.02; **HRMS (EI)** calcd m/z for $C_{13}H_{13}ClIO$: 346.9695 [M$^+$-CH$_3$] found 346.9695; **Elemental Analysis** calcd (%) for $C_{14}H_{16}ClIO$: C 46.37, H 4.45, Cl 9.78, I 34.99; found: C 46.44, H 4.54, Cl 10.07, I 34.77, O 4.51; **CAS**: 1240913-25-0**CCDC**: 771243.

1-Trifluoromethyl-3,3-dimethyl-3a,6-methano-3a,4,5,6-tetrahydro-1H,3H-λ^3-ioda-2-oxa-phenalene (29)

Dry KOAc (0.337 g, 3.4 mmol, 1.7 equiv) and **16**(0.733 g, 2.0 mmol) were suspended in CH_3CN (5 mL). After 4h the suspension was filtered, washed with CH_3CN (2x 2.5 mL) and the filtrate was cooled to –40 °C. TMSCF$_3$ (0.45 mL, 3.0 mmol, 1.5 equiv) followed by TBAT (0.2 mL, 0.03 M in CH_3CN, 0.3 mol-%) was added and the mixture was stirred over night (16 h). The mixture was subsequently allowed to warm to 0 °C (5 °C/1h), after 2 h and 4 h additional TMSCF$_3$ (67 µL, 0.45 mmol, 0.22 equiv) was added, when the temperature reached 0 °C TMSCF$_3$ (0.15 mL, 1 mmol, 0.5 equiv) was added and the mixture was allowed to warm to room temperature where it was stirred for 3 h. The solvent was removed under reduced pressure, after purification by flash chromatography (SiO$_2$, pentane/Et$_2$O 2:1) **29** (114 mg, 14%) was isolated as a microcrystalline solid. R_f (pentane/Et$_2$O 2:1): 0.25; ^{1}**H NMR** (500 MHz, CDCl$_3$): 7.48 (d, $J_{H,H}$ = 8.5 Hz, 1H, C$_{Ar}H$), 7.29 (d, $J_{H,H}$ = 7.0, 1H, C$_{Ar}H$), 7.17 (dd, $J_{H,H}$ = 8.5 Hz, $J_{H,H}$ = 7.0 Hz, 1H, C$_{Ar}H$), 3.36 (u, 1H, CH), 2.05 (m, 2H, CHH$_{eq}$CHH$_{eq}$), 1.67 (dq, $J_{H,H}$ = 9.0 Hz, $J_{H,H}$ = 2.0 Hz, 1H, CH$_{2,endo}$), 1.54 (dd, $J_{H,H}$ = 9.0 Hz, $J_{H,H}$ = 1.3, 1H, CH$_{2,exo}$), 1.49 (s, 3H, CH$_3$), 1.38 (s, 3H, CH$_3$), 1.37-1.21 (m, 2H, $CH_{ax}HCH_{ax}H$); $^{13}C\{^1H\}$ **NMR** (75.5 MHz, CDCl$_3$): δ = 154.8, 149.9, 129.1, 125.3, 122.6, 116.6(q, $J_{C,F}$ = 493 Hz, CF$_3$), 110.9(q, $^3J_{C,F}$ =3.2 Hz, CI), 72.4, 58.9, 49.0, 42.6, 29.3, 28.7, 26.6 (2C); ^{19}**F NMR** (188 MHz, CDCl$_3$): δ = –44.2(s, $^1J_{C,F}$ = 403 Hz, CF$_3$); **HRMS (EI)** calcd m/z for $C_{14}H_{13}F_3IO$: 380.9958 [M$^+$-CH$_3$], found: 380.9962 [M$^+$-CH$_3$]; **Elemental Analysis** calcd (%) for $C_{15}H_{16}F_3IO$: C 45.47, H 4.07, O 4.04, F 14.39, I 32.03; found: C 45.47, H 4.09, F 14.34, I 32.29; **CAS**: 1240913-43-2; **CCDC**: 771237.

7-Chloro-5,5-dimethyl-7λ^3-ioda-3-oxa-6λ^5-azatricyclo[6.4.0.0^{2,6}]dodeca-1(8),2(6),9,11-tetraen-6-ylium tetrafluoro-λ^4-borane (20)

Oxazolidine **22** (485 mg, 1.61 mmol) was dissolved in Et$_2$O (6 mL). After the addition of HBF$_4$-Et$_2$O (0.7 mL, 51-57% in Et$_2$O, 2.41 mmol, 1.5 equiv) a white precipitate was formed, which was filtered after 90 min, washed with Et$_2$O and dissolved in CH$_2$Cl$_2$ (14 mL). In the dark, tBuOCl (0.33 mL, 3.22 mmol, 2.0 equiv) was added to the resulting solution dropwise. A white precipitate formed immediately. After

Experimental Part

the suspension was stirred for an additional 5 h, all volatile compounds were removed under reduced pressure. The white solid was washed with Et_2O and dried under HV to yield compound **20** (542 mg, 80%). X-ray quality single crystals (hygroscopic plate) were obtained by diffusion of CH_2Cl_2 into a saturated acetonitrile solution. **^{1}H NMR** (250 MHz, CD_3CN): δ = 8.48 (d, $J_{H,H}$ = 8.8, 1H, $C_{Ar}H$), 8.20 (m, 2H, $C_{Ar}H$), 8.03 (t, $J_{H,H}$ = 7.5 Hz, 1H, $C_{Ar}H$), 4.94 (s, 2H, CH_2), 1.58 (s, 6H, 2x CH_3); **$^{13}C\{^1H\}$ NMR** (62.9 MHz, CD_3CN): δ = 171.2, 138.1, 132.6, 131.3, 129.4, 123.2, 122.0, 86.0, 69.0, 26.6; **^{19}F NMR** (188 MHz, CD_3CN): −151.3; **HRMS (MALDI)** calcd m/z for $C_{11}H_{12}ClINO$ 335.9647 [M$^+$], found 335.9644 [M$^+$]; **Elemental Analysis** calcd (%) for $C_{11}H_{12}BClF_4INO$: C 31.21, H 2.86, B 2.55, Cl 8.37, F 17.95, I 29.97, N 3.31, O 3.78; found: C 31.11, H 2.84, N 3.28; **CAS**: 1240913-31-8; **CCDC**: 771242.

8-Chloro-8λ^3-ioda-7λ^5-azatricyclo-[7.4.0.0^{2,7}]trideca-1(9),2,4,6,10,12-hexaen-7-ylium tetrafluoro-λ^4-borane (21)

2-(2-iodophenyl)pyridine (**23**) (205 mg, 0.73 mmol) was dissolved in Et_2O (2.5 mL). After dropwise addition of HBF_4-Et_2O (0.31 mL, 51-57% in Et_2O, 1.07 mmol, 1.5 equiv), a pale brown precipitate formed. The solvent was decanted by syringe after 90 min and the residual solid was washed once with Et_2O (2.5 mL), then dried under HV. CH_2Cl_2 (6.0 mL) was added yielding a clear solution. Dropwise addition of tBuOCl (0.16 mL, 0.15 mmol, 2.0 equiv) in the dark at ambient temperature yielded a white precipitate. The suspension was stirred for additional 12 h, then all volatiles were removed at HV. The residual solid was washed with CH_2Cl_2 and dried under HV to yield compound **21** (257 mg, 0.64 mmol, 87%) as a pale yellow solid. X-ray quality single crystals (hygroscopic plate) were obtained by diffusion of CH_2Cl_2 into a saturated acetonitrile solution. **^{1}H NMR** (300 MHz, CD_3CN,): δ = 9.04 (d, $J_{H,H}$ = 5.9 Hz, 1H), 8.64-8.53 (m, 3H), 8.40 (d, $J_{H,H}$ = 7.9 Hz, 1H), 8.06-8.02 (m, 2H), 7.90 (td, $J_{H,H}$ = 6.6 Hz, $J_{H,H}$ = 1.3 Hz, 1H); **$^{13}C\{^1H\}$ NMR** (75 MHz, CD_3CN): δ = 150.4, 144.3, 143.3, 135.6, 132.6, 132.3, 129.6, 129.1, 127.5, 123.7, 115.0; **^{19}F NMR** (188 MHz, CD_3CN): −150.8; **HRMS (EI)** calcd m/z for $C_{11}H_8ClIN$: 315.9484 [M$^+$], found: 315.9383 [M$^+$]; **Elemental Analysis** calcd (%) for $C_{11}H_8NBF_4ClI$: C 32.76, H 2.00, N 3.47; found: C 32.68, H 2.11, N 3.44; **CAS**: 1240913-36-3; **CCDC** 771238.

1-(2-Fluoropropan-2-yl)-2-iodobenzene (26a)

The reaction was carried out in a Teflon® test-tube under an inert atmosphere. To a solution of 2-(2-iodophenyl)propan-2-ol (**4a**) (2.05 g, 7.82

Experimental Part

mmol) in CH_2Cl_2 (1.5 mL) was added to Deoxo-Fluor™ (1.72 mL, 9.33 mmol, 1.2 equiv) in CH_2Cl_2 (2.2 mL) at –78 °C under stirring. The reaction mixture was stirred at –78 °C for 1 h and at ambient temperature for an additional hour. The mixture was poured into saturated aqueous NH_4Cl (50 mL) and the aqueous phase was extracted with CH_2Cl_2 (3x 40 mL). The combined organic phases were dried over $MgSO_4$, filtered and the solvent removed under vacuum. The residue was purified by flash chromatography (SiO_2, pentane) to yield the title compound as a yellow liquid (1.338 g, 70%). R_f (pentane): 0.52; **^{1}H NMR** (300 MHz, $CDCl_3$): δ = 7.98 (d, $J_{H,H}$ = 7.8 Hz, 1H, $C_{Ar}H$), 7.61 (dd, $J_{H,H}$ = 7.8 Hz, $J_{H,H}$ = 1.5 Hz, 1H, $C_{Ar}H$), 7.38 (dt, $J_{H,H}$ = 7.5 H, $J_{H,H}$ = 0.9 Hz, 1H, $C_{Ar}H$), 6.95 (dt, $J_{H,H}$ = 7.8 Hz, $J_{H,H}$ = 1.8 Hz, 1H, $C_{Ar}H$), 1.9 (d, $J_{H,H}$ = 23.7 Hz, 6H, 2x CH_3); **^{13}C{^{1}H} NMR** (125.8 MHz, CD_2Cl_2): δ = 147.4 (d, $J_{C,F}$ = 21.3 Hz, C_{Ar}), 142.8, 129.4, 128.6 (d, $J_{C,F}$ = 2.9 Hz, C_{Ar}), 126.2 (d, $J_{C,F}$ = 18.5 Hz, C_{Ar}), 96.8 (d, $^1J_{C,F}$ = 171.5 Hz, CF), 27.7 (d, $J_{C,F}$ = 25.5, 2C, 2x CH_3); **^{19}F NMR** (282 MHz, $CDCl_3$): δ = –128.3 (hept, $J_{F,H}$ = 23.4 Hz, CF); **HRMS (EI)** cald m/z for C_8H_7FI: 248.9571 $[M^+-CH_3]$, found: 248.9568 $[M^+-CH_3]$; **Elemental Analysis** calcd (%) for $C_9H_{10}FI$: C 40.93, H 3.82, F 7.19, I 48.06; found: C 40.94, H 3.75, F 7.18, I 47.82.

1-(Dichloro-λ^3-iodanyl)-2-(1-fluoro-1-methylethyl)benzene (25a)

26a (80 µL, 0.5 mmol) was dissolved in hexane (6 mL) in a round-bottomed Young Schlenk flask (25 mL). In the dark, Cl_2 was bubbled through the solution for 1.5 h. The mixture was washed with pentane (2x 5 mL) and filtered. The yellow precipitate was dried under vacuum to yield the title compound as yellow crystals (124 mg, 74%). The product decomposes in CH_3CN and under vacuum. Crystals for X-Ray analysis were obtained by diffusing pentane into a saturated CH_2Cl_2 solution. **^{1}H NMR** (200 MHz, $CDCl_3$): δ = 8.51 (dd, $J_{H,H}$ = 8.0 Hz, $J_{H,H}$ = 1.0 Hz, 1H, $C_{Ar}H$), 7.69 (tt, $J_{H,H}$ = 7.8 Hz, $J_{H,H}$ = 1.2 Hz, 1H $C_{Ar}H$), 7.51 (dt, $J_{H,H}$ = 8.0, $J_{H,H}$ = 1.6 Hz, 1H, $C_{Ar}H$), 7.39 (td, $J_{H,H}$ = 8.0 Hz, $J_{H,H}$ = 1.8 Hz, 1H, $C_{Ar}H$), 1.98 (d, $J_{H,F}$ = 22.4 Hz, 6H, 2x CH_3). **^{19}F NMR** (282 MHz, $CDCl_3$): δ = –135.7 (hept, $J_{F,H}$ = 22.4 Hz, CF). **Elemental Analysis** calcd (%) for $C_9H_{10}Cl_2FI$: C 32.27, H 3.01, F 5.67, Cl 21.17, I 37.88; found: C 32.49, H 2.98, F 5.49, Cl 21.46.

1-(Dichloro-λ^3)-iodanyl)-2-trifluoromethylbenzene (25b)

2-Iodobenzotrifluoroide (250 mL, 1.77 mmol) was dissolved in glacial acetic acid (3.5 mL). $Na_2S_2O_8$ (842 mg, 3.54 mmol, 2 equiv) was added to the stirred solution. HCl (aq. 37%, 1.5 mL, 17.7 mmol, 10 equiv) was added dropwise under stirring. The reaction mixture was warmed to 45 °C and stirred for 16 h. The mixture was poured into ice-water while stirring. The yellow precipitate was filtered and washed with ice-cold water until the pH was neutral, followed by pentane (2x 5 mL) and dried under vacuum to yield the title compound as yellow crystals (512 mg, 85%). The product decomposes in CH_3CN and under vacuum. ^{1}**H NMR** (300 MHz, CD_2Cl_2): δ = 8.58 (d, $J_{H,H}$ = 8.1 Hz, 1H, $C_{Ar}H$), 7.96 (d, $J_{H,H}$ = 8.1 Hz, 1H, $C_{Ar}H$), 7.88 (t, $J_{H,H}$ = 7.5 Hz, 1H, $C_{Ar}H$), 7.74 (t, $J_{H,H}$ = 7.8 Hz, $C_{Ar}H$); ^{1}**H NMR** (700 MHz, SO_2): δ = 8.96 (d, 1H, C^6H), 8.39 (d, 1H, C^3H), 8.32 (t, 1H, C^4H), 8.19 (t, 1H, C^5H); ^{13}**C (HMQC) NMR** (126 MHz; SO_2): δ = 140.0 (1C, C^6), 137.1 (1C, C^5), 134.8 (1C, C^4), 129.4 (1C, C^3), 129.1 (1C, C^2), 122.9 (1C, C^7), 119.6 (1C, C^1); ^{19}**F NMR** (282 MHz, CD_2Cl_2): δ = –60.4 (s, CF_3).

1-Fluoro-1,3-dihydro-3,3-dimethyl-1,2-benziodoxole (30a)

Method A: TFMT (427 mg, 2.0 mmol, 1.3 equiv) was condensed into a Young Schlenk (10 mL) at -30 °C. After the addition of acetonitrile (6 mL), AgF (229 mg, 1.8 mmol, 1.2 equiv) was added in one portion. In the dark, the reaction mixture was allowed to warm to 0 °C over 2 h. The mixture was then cooled to –30 °C and the CF_3SO_2F formed during the reaction, was removed under HV (15 min). In a separate Schlenk, 1-chloro-1,3-dihydro-3,3-dimethyl-1,2-benziodoxole (**2a**) (447 mg, 1.5 mmol) was dissolved in CH_2Cl_2 (6 mL) and cooled to –78 °C. The $AgOCF_3$ solution was added and the mixture was allowed to warm to room temperature. All volatile compounds were removed under reduced pressure and the residue was taken up in CH_2Cl_2 (3 mL), filtered and the solvent removed under reduced pressure to yield the title compound as a colorless powder (407 mg, 97%).

Method B: In a Young Schlenk (10 mL) KF (116 mg, 2.0 mmol, 2.0 equiv) and 1-chloro-1,3-dihydro-3,3-dimethyl-1,2-benziodoxole (**2a**) (302 mg, 1.0 mmol) were suspended in CH_3CN (3 mL) and stirred for two hours at room temperature. Afterwards, the colorless suspension was filtered and the solvent removed under reduced pressure. The solid was washed with cold pentane (2x 1 mL) and dried to yield the titled compound as with powder (285 mg, quant, purity > 92%). Analytic samples were obtained after sublimation (35 °C, 10^{-3}mbar).

^{1}**H NMR** (400.1 MHz, CD_3CN): δ = 7.73 (dd, $J_{H,H}$ = 8 Hz, 0.8 Hz, 1H, C^6H), 7.61 (dd, $J_{H,H}$ = 8 Hz, $J_{H,H}$ = 1.5 Hz, 1H, C^5H), 7.55 (dd, $J_{H,H}$ = 7.5 Hz, $J_{H,H}$ = 1.2 Hz, 1H, C^4H), 7.32 (dd, $J_{H,H}$ = 7.5 Hz, $J_{H,H}$ = 1.2 Hz, 1H, C^3H), 1.49 (s, 6H, 2x CH_3), ^{13}**C{^{1}H} NMR** (75.5 Hz, CD_3CN): δ = 149.1 (C^2), 130.6 (C^4H), 130.1 (C^5H), 128.1 (d, $J_{C,F}$ = 8.5 Hz, C^6H), 126.2 (C^3H), 115.6 (d, $J_{C,F}$ = 8.8 Hz, CI), 84.7 (C(CH)$_3$), 28.3 (u d, $J_{C,F}$ = 2.3 Hz, 2x CH_3); ^{19}**F NMR** (376.5 Hz, CD_3CN): δ = –140.8; **HRMS (EI)** calcd m/z for $C_9H_{10}FIO$: 279.9755 [M$^+$], 264.9521 [M$^+$-CH$_3$], found: 279.9731 [M$^+$, 0.7%], 264.9522 [M$^+$-CH$_3$, 100%];

Experimental Part

Elemental Analysis calcd (%) for $C_9H_{10}FIO$: C 38.60, H 3.60, F 6.78, I 45.31, O 5.71; found: C 38.66, H 3.65.

1-Fluoro-1,3-dihydro-3,3-bis(trifluoromethyl)-1,2-benziodoxole (30b)

Method A: TFMT (244 mg, 1.12 mmol, 1.6 equiv) was condensed into a Young Schlenk (10 mL) at −30 °C. After the addition of acetonitrile (4 mL), AgF (129 mg, 1.5 equiv) was added in one portion. In the dark, the reaction mixture was allowed to warm to 0 °C over 2 h. The mixture was then cooled to −30 °C and the CF_3SO_2F formed during the reaction, was removed under HV (10 min). In a separate Schlenk, 1-chloro-1,3-dihydro-3,3-bis(trifluoromethyl)-1,2-benziodoxole (**2c**) (274 mg, 0.067 mmol) was dissolved in CH_2Cl_2 (4 mL) and cooled to −78 °C. The $AgOCF_3$ solution was added and the mixture was allowed to warm to −30 °C. After 1 h, the reaction mixture was filtered and all volatile compounds were removed under reduced pressure to yield the title compound as a colorless powder (239 mg, 91%).

Method B: In a Young Schlenk (10 mL) KF (116 mg, 2.0 mmol, 2.0 equiv) and 1-chloro-1,3-dihydro-3,3-bis(trifluoromethyl)-1,2-benziodoxole (**2c**) (302 mg, 1.0 mmol) were suspended in CH_3CN (3 mL) and stirred for 6 days at room temperature. Afterwards, the colorless suspension was filtered and the solvent removed under reduced pressure. The solid was washed with cold pentane (2x 1 mL) and further purified by sublimation (35 °C, 10^{-3} mbar) to yield the titled compound as colorless microcrystals (170 mg, 44%). **30b** slowly decomposes in solution.

^{1}H NMR (700.1 MHz, CD_3CN): δ = 7.97 (dd, $J_{H,H}$ = 7.1 Hz, $J_{H,H}$ = 1.8 Hz 1H, C^4H), 7.89 (dm, $J_{H,H}$ = 8.9 Hz, 1H, CH), 7.80 (dm, $J_{H,H}$ = 8.0 Hz, 1H, C^3H), 7.78 (m, C^5H); **$^{13}C\{^1H\}$ NMR** (176.1 Hz, CD_3CN): δ = 134.2 (1C, C^4), 131.5 (1C, C^5H), 130.4 (1C, C^2H), 129.3 (1C, C^3H), 128.3 (d, $J_{C,F}$ = 11 Hz, C^6H), 123.2 (q, $J_{C,F}$ = 287 Hz, 2C, 2x CF_3), 116.9 (d, $J_{C,F}$ = 10 Hz, CI), 85.8 (hept, $J_{C,F}$ = 30 Hz, 1C, $C(CF_3)_2$), **^{19}F NMR** (376.5 Hz, CD_3CN): δ = −76.6 (6F, 2x CF_3), −173.4 (1F, FI); **HRMS (EI)** calcd m/z for $C_9H_{10}FIO$: 387.9190 [M$^+$], 368.9206 [M$^+$-I], 318.9238 [M$^+$-CF_3], found: 387.9205 [M$^+$, 2.7%], 368.9208 [M$^+$-I, 3.8%], 318.9237 [M$^+$-CF_3, 100%]; **Elemental Analysis** calcd (%) for $C_9H_4F_7IO$: C 27.86, H 1.04, F 34.27, I 32.71, O 4.12; found: C 27.73, H 1.19.

Experimental Part

5.3 Nitrogen-Center Nucleophiles

5.3.1 Ritter-type Reaction

General procedure for Heterocyclic N-substituted N-trifluoroimines

A flame-dried Young Schlenk (20 mL) was charged with 1-trifluoromethyl-1,3-dihydro-3,3-dimethyl-1,2-benziodoxole (**1a**) (198 mg, 0.60 mmol) and azole (0.90 mmol, 1.5 equiv). CH$_3$CN (6 mL) and HNTf$_2$ (0.6 mL, 0.1 M in CH$_2$Cl$_2$, 10 mol-%) were added. The mixture was stirred at 60 °C for 3.5 h. The mixture was extracted with pentane (3 x 15 mL) and the pentane was removed under reduced pressure.

(*E*)-*N*-(1-(1*H*-Benzo[*d*][1,2,3]triazol-1-yl)ethylidene)trifluoromethanamine (39)

The title compound was synthesized from benzotriazole (107 mg, 0.90 mmol, 1.5 equiv) according to the general procedure and purified by flash chromatography (SiO$_2$, pentane/CH$_2$Cl$_2$ 5:1) to yield **39** (86.5 mg, 63%) as white crystals. Single crystals for X-Ray analysis were obtained by slow evaporation of CHCl$_3$. **m.p.**: 91 °C; **R_f** (pentane/CH$_2$Cl$_2$ 5:1): 0.45; **^{1}H NMR** (500.2 MHz, CDCl$_3$): δ = 8.41 (d, $J_{H,H}$ = 8.3 Hz, 1H, C^7*H*), 8.13 (d, $J_{H,H}$ = 8.3 Hz, 1H, C^4*H*), 7.66 (t, $J_{H,H}$ = 7.7 Hz, 1H, C^6*H*), 7.52 (t, $J_{H,H}$ = 7.7 Hz, 1H, C^5*H*), 3.13 (br q, $J_{H,F}$ = 0.8 Hz, 3H, C*H*$_3$); **^{13}C{^{1}H} NMR** (125.8 MHz, CDCl$_3$): δ = 164.9 (q, $J_{C,F}$ = 7.7 Hz, 1C, N^1*C*CH$_3$), 147.1 (1C, *C*^{3a}), 131.1 (1C, *C*^{7a}), 130.7 (1C, *C*^{6}H), 126.6 (1C, *C*^{5}H), 124.0 (q, $^1J_{C,F}$ = 260 Hz, *C*F$_3$), 120.5 (1C, *C*^{4}H), 116.2 (1C, *C*^{7}H), 18.4 (q, $J_{C,F}$ = 1.8 Hz, *C*H$_3$); **^{15}N NMR** (40.6 MHz, CDCl$_3$): δ = 254.5 (1N, *N*CF$_3$), 245.3 (1N, *N*^{1}), *N*^{2} and *N*^{3} not observed; **^{19}F NMR** (376.5 MHz, CDCl$_3$): δ = −53.6 (br q, $J_{F,H}$ = 1.1 Hz, 3F, C*F*$_3$); **HRMS (EI)** calcd m/z for C$_9$H$_7$F$_3$N$_4$: 228.0623 [M$^+$], found: 228.0618 [M$^+$]; **Elemental Analysis** calcd (%) for C$_9$H$_7$F$_3$N$_4$: C 47.38, H 3.09, F 24.98, N 24.55, found: C 47.42, H 3.24, F 24.97, N 24.49; **CAS**: 1269630-36-5; **CCDC**: 792179.

(*E*)-*N*-(1-(1*H*-Benzo[*d*][1,2,3]triazol-1-yl)propylidene)-1,1,1-trifluoromethanamine (50)

The title compound was prepared from benzotriazole (107 mg, 0.90 mmol, 1.5 equiv) according to the general procedure using EtCN (6 mL) instead of CH$_3$CN. After the reaction the solvent was remove under reduced pressure and the residue was purified by flash chromatography (SiO$_2$, pentane/CH$_2$Cl$_2$ 5:1) to yield corresponding imine (53.9 mg, 37%) as white crystals. **m.p.**: 73 °C; **R_f** (pentane/CH$_2$Cl$_2$ 5:1): 0.22; **^{1}H NMR** (700.1 MHz, CDCl$_3$):

Experimental Part

δ = 8.40 (d, $J_{H,H}$ = 8.4 Hz , 1H, C^7H), 8.12 (d, $J_{H,H}$ = 8.3 Hz, 1H, C^4H), 7.65 (ddd, $J_{H,H}$ = 8.4 Hz, $J_{H,H}$ = 7.6 Hz, $J_{H,H}$ = 0.9 Hz, 1H, C^6H), 7.52 (ddd, $J_{H,H}$ = 8.2 Hz, $J_{H,H}$ = 7.1 Hz, $J_{H,H}$ = 1.0 Hz, 1H, C^5H), 3.47 (q, $J_{H,H}$ = 7.6 Hz, 2H, CH_2), 1.50 (t, $J_{H,H}$ = 7.6 Hz, 1H, CH_3); $^{13}C\{^1H\}$ **NMR** (176.0 MHz, $CDCl_3$): δ = 169.4 (q, $^3J_{C,F}$ = 7.0 Hz, 1C, N^1CCH_2), 146.9 (1C, C^{3a}), 131.2 (1C, C^{7a}), 130.6 (1C, C^6H), 126.5 (1C, C^5H), 124.0 (q, $^1J_{C,F}$ = 260 Hz, 1C, CF_3), 120.4 (1C, C^4H), 116.2 (1C, C^7H), 25.8 (br q, $J_{C,F}$ = 1.4 Hz, 1C, CH_2), 13.0 (1C, CH_3); ^{19}F **NMR** (376.5 MHz, $CDCl_3$): δ = –52.9 (s, 3F, CF_3); **HRMS (EI)** calcd m/z for $C_{10}H_9F_3N_4$: 242.0774 [M$^+$], found: 242.0775 [M$^+$]; **Elemental Analysis** calcd (%) for $C_{10}H_9F_3$ N_4: C 49.59, H 3.75, F 23.53, N 23.13, found: C 49.93, H 3.96, F 23.37, N 22.80; **CAS**: 1269630-37-6.

(E)-N-(1-(1H-Benzo[d][1,2,3]triazol-1-yl)propylidene)-1,1,1-trifluoromethanamine (51)

The title compound was prepared from benzotriazole (107 mg, 0.90 mmol, 1.5 equiv) according to the general procedure using iPrCN (6 mL) instead of CH_3CN and purified by flash chromatography (SiO_2, pentane/CH_2Cl_2 10:1) to yield corresponding imine (91 mg, 36%) as white crystals. **m.p.**: 72-73 °C; R_f (pentane/CH_2Cl_2 10:1): 0.18; ^1H **NMR** (300.1 MHz, $CDCl_3$): δ = 8.41 (Ψdt, $J_{H,H}$ = 8.4 Hz, $J_{H,H}$ = 1 Hz, 1H, C^7H), 8.14 (Ψdt, $J_{H,H}$ = 8.2 Hz, $J_{H,H}$ = 1 Hz, 1H, C^4H), 7.67 (dd, $J_{H,H}$ = 7.2 Hz, $J_{H,H}$ = 1.2 Hz, 1H, C^6H), 7.54 (dd, $J_{H,H}$ = 7.1 Hz, $J_{H,H}$ = 1.1 Hz, 1H, C^5H), 3.81 (hept, $J_{H,H}$ = 7 Hz, 1H, CH_{iPr}), 1.71 (d, $J_{H,H}$ = 7 Hz, 6H, 2x CH_3); $^{13}C\{^1H\}$ **NMR** (75.5 MHz, $CDCl_3$): δ = 171.7 (q, $^3J_{C,F}$ = 6.7 Hz, 1C, $CNCF_3$), 145.7 (1C, C^{3a}), 131.6 (1C, C^{7a}), 130.4 (1C, C^6H), 126.3 (1C, C^5H), 123.7 (q, $^1J_{C,F}$ = 260 Hz, 1C, CF_3), 120.1 (1C, C^4H), 116.4 (1C, C^7H), 34.5 (u q, $J_{C,F}$ = 1.6 Hz, 1C, CH_{iPr}), 20.0 (1C, 2x CH_3); ^{19}F **NMR** (282.4 MHz, $CDCl_3$): δ = –52.2 (s, 3F, CF_3); **HRMS (EI)** calcd m/z for $C_{11}H_{11}F_3N_4$ 256.0931 [M$^+$], found: 256.0930 [M$^+$, 3.2%], 213.0643 [M$^+$-N_2-CH_3, 100%]; **Elemental Analysis** calcd (%) for $C_{11}H_{11}F_3N_4$: C 51.56, H 4.33, F 21.87, N 22.24, found: C 51.59, H 4.47, F 21.87, N 22.49.

(E)-N-(1-(1H-Benzo[d][1,2,3]triazol-1-yl)-2-phenylethylidene)-1,1,1-trifluoromethan-amine (52)

The title compound was prepared from benzotriazole (107 mg, 0.90 mmol, 1.5 equiv) according to the general procedure using PhCN (6 mL) instead of CH_3CN and the solvent was removed under reduced pressure (10^{-3} mbar) after extraction. The residue was purified by flash chromatography (SiO_2, pentane/CH_2Cl_2 1:1) to yield corresponding imine (25 mg, 14%) as white crystals. **m.p.**: 133-135 °C; R_f (pentane/CH_2Cl_2 1:1): 0.43; ^1H **NMR** (400.1 MHz, $CDCl_3$): δ = 8.53 (dd, $J_{H,H}$ = 8.4 Hz, $J_{H,H}$ = 0.8 Hz, 1H, C^7H), 8.14

(dd, $J_{H,H}$ = 8.0 Hz, $J_{H,H}$ = 0.8 Hz, 1H, C^4H), 7.67 (u ddd, $J_{H,H}$ = 7.3 Hz, $J_{H,H}$ = 0.8 Hz, $J_{H,H}$ = 0.8 Hz, 1H, C^6H), 7.69-7.56 (m, 6H, C^5H and 5x CH_{Ph}); $^{13}C\{^1H\}$ **NMR** (100.6 MHz, CDCl$_3$): δ = 164.1 (q, $J_{C,F}$ = 7.7 Hz, 1C, $CNCF_3$), 147.2 (1C, C^{3a}), 131.9 (1C, $C_{ipso-Ph}$), 131.8 (1C, CH_{p-Ph}), 130.9 (1C, C^6H), 130.7 (1C, C^{7a}), 128.9 (2C, CH_{o-Ph}), 128.6 (2C, CH_{m-Ph}), 127.1 (1C, C^5H), 126.0 (q, $^1J_{C,F}$ = 224 Hz, 1C, CF_3), 122.3 (1C, C^4H), 116.4 (1C, C^7H); ^{19}**F NMR** (282.4 MHz, CDCl$_3$): δ = −52.2 (s, 3F, CF_3); **HRMS (EI)** calcd m/z for $C_{14}H_9F_3N_4$: 290.0774 [M$^+$], found: 290.0775 [M$^+$]; **Elemental Analysis** calcd (%) for $C_{14}H_9F_3N_4$: C 57.93, H 3.13, F 19.64, N 19.30, found: C 58.06, H 3.32, F 19.56, N 19.13.

(*E*)-*N*-(1-(1*H*-Benzo[*d*][1,2,3]triazol-1-yl)(phenyl)methylene)-1,1,1-trifluoromethanamine (53)

The title compound was prepared from benzotriazole (107 mg, 0.90 mmol, 1.5 equiv) according to the general procedure using BnCN (6 mL) instead of CH$_3$CN and purified by flash chromatography (SiO$_2$, pentane/Et$_2$O 10/1), the solvent of the product containing fractions was removed under reduced pressure, and the residue washed with cold pentane to yield the title compound (13 mg, 7%) as colorless powder. R_f (pentane/Et$_2$O 10:1): 0.5; ^{1}**H NMR** (300.1 MHz, CDCl$_3$): δ = 8.45 (d, $J_{H,H}$ = 8.4 Hz, 1H, C^7H), 8.12 (d, $J_{H,H}$ = 8.1 Hz, 1H, C^4H), 7.69 (dd, $J_{H,H}$ = 7.2 Hz, $J_{H,H}$ = 0.9 Hz, 1H, C^6H), 7.54 (dd, $J_{H,H}$ = 7.1 Hz, $J_{H,H}$ = 1.2 Hz, 1H, C^5H), 7.38-7.24 (m, 5H, CH_{Ph}), 4.89 (s, 2H, CH_2); $^{13}C\{^1H\}$ **NMR** (75.5 MHz, CDCl$_3$): δ = 164.6 (q, $^3J_{C,F}$ = 6.9 Hz, 1C, $CNCF_3$), 146.8 (1C, C^{3a}), 133.6 (1C, C^{7a}), 131.14 (1C, $C_{ipso-Ph}$), 130.5 (1C, C^6H), 128.8 (2C, 2x CH_{m-Ph}), 128.7 (br s, 2C, CH_{o-Ph}), 127.4 (1C, CH_{p-Ph}), 126.5 (1C, C^5H), 123.6 (q, $^1J_{C,F}$ = 260 Hz, 1C, CF_3), 120.3 (1C, C^4H), 116.0 (1C, C^7H), 37.5 (q, $J_{C,F}$ = 1.6 Hz, C, CH_2), 20.0 (1C, 2x CH_3); ^{19}**F NMR** (282.4 MHz, CDCl$_3$): δ = −51.9 (s, 3F, CF_3); **HRMS (EI)** calcd m/z for $C_{15}H_{11}F_3N_4$: 304.0931 [M$^+$], found: 304.0930 [M$^+$]; **Elemental Analysis** calcd (%) for $C_{15}H_{11}F_3N_4$: C 59.21, H 3.64, F 18.73, N 18.73, found: C 59.23, H 3.64, F 18.90, N 18.17.

(*E*)-*N*-(1-(2*H*-Indazol-1-yl)ethylidene)-1,1,1-trifluoromethanamine (54a)

The title compound was prepared from indazole (106 mg, 0.9 mmol, 1.5 equiv) according to the general procedure and filtered over a silica pad with pentane/CH$_2$Cl$_2$ = 2:1. The crude product was purified by sublimation (r.t., 2 10^{-2} mbar) to give **54a** (63.2 mg, 47%) as white crystals. Single crystals for X-ray analysis were obtained by slow evaporation of a CHCl$_3$ solution. **m.p.**: 121 °C; ^{1}**H NMR** (400.1 MHz, CDCl$_3$): δ = 8.90 (d, $J_{H,H}$ = 1 Hz, 1H, C^3H), 7.64 (ddd, $J_{H,H}$ = 9.0 Hz, $J_{H,H}$ = 2 Hz, $J_{H,H}$ = 1 Hz, 1H, C^7H), 7.61 (Ψdt, $J_{H,H}$ = 8.7 Hz, $J_{H,H}$ = 1

Experimental Part

Hz, 1H, C^4H), 7.32 (ddd, $J_{H,H}$ = 9 Hz, .$J_{H,H}$ = 7.5 Hz, $J_{H,H}$ = 1 Hz, 1H, C^6H), 7.09 (ddd, $J_{H,H}$ = 8.7 Hz, $J_{H,H}$ = 6.5 Hz, $J_{H,H}$ = 0.6 Hz, 1H, C^5H), 3.00 (br q, $J_{H,F}$ = 1.2 Hz, 3H, CH_3); $^{13}C\{^1H\}$ NMR (100.6 MHz, $CDCl_3$): δ = 166.2 (q, $J_{C,F}$ = 7.2 Hz, 1C, CCH_3), 151.5 (1C, C^{7a}), 129.4 (1C, C^6), 124.6 (1C, C^5), 123.7 (q, $^1J(C,F)$= 260 Hz, 1C, CF_3), 122.9 (1C, C^{3a}), 122.0 (1C, C^3), 121.5 (1C, C^4), 118.8 (1C, C^7), 17.5 (q, $J(C,F)$= 1.7 Hz, 1C, CH_3); ^{15}N NMR (40.6 MHz, $CDCl_3$): δ = 236.5 (N^2), 254.1 (NCF_3), N^1 not observed; ^{19}F NMR (376.5 MHz, $CDCl_3$): δ = −54.0 (s, 3F, CF_3); **HRMS (EI)** calcd m/z for $C_{10}H_8F_3N_3$: 227.0665 [M$^+$], found: 227.0668 [M$^+$]; **Elemental Analysis** calcd (%) for $C_{10}H_8N_3F_3$: C 52.87, H 3.55, F 25.09, N 18.50, found: C 52.83, H 3.72, F 25.08, N 18.44; **CAS**: 1269630-38-7; **CCDC**: 792180.

(*E*)-*N*-(1-(1*H*-Indazol-1-yl)ethylidene)-1,1,1-trifluoromethanamine (54b)

54a (50 mg, 0.22 mmol) was dissolved in CH_3CN (2 mL), $HNTf_2$ (0.1 M in CH_2Cl_2, 0.22 mL, 22 µmol, 10 mol-%) was added and the mixture was stirred at 70 °C for 24 h and subsequentlyextracted with pentane. The pentane was removed under reduced pressure and the resulting white powder was purified by sublimation (r.t., 10^{-2} mbar) to yield regioisomer **54b** (30.1 mg, 60%) as white crystals. Single crystals for X-ray analysis were obtained by sublimation (room temperature, 10^{-2} mbar). **m.p.**: 72 °C; ^1H **NMR** (400.1 MHz, $CDCl_3$): δ = 8.65 (ddd, $J_{H,H}$ = 8.5 Hz, $J_{H,H}$ = 1.8 Hz, $J_{H,H}$ = 0.9 Hz, 1H, C^7H), 8.19 (s, 1H, C^3H), 7.75 (ddd, $J_{H,H}$ = 8.4Hz, $J_{H,H}$ = 7.5 Hz, $J_{H,H}$ = 0.7 Hz, 1H, C^4H), 7.58 (ddd, $J_{H,H}$ = 8.4 Hz, $J_{H,H}$ = 7.1 Hz, $J_{H,H}$ = 1.2 Hz, 1H, C^6H), 7.38 (ddd, $J_{H,H}$ = 8.0 Hz, $J_{H,H}$ = 7.2 Hz, $J_{H,H}$ = 0.9 Hz, 1H, C^5H), 2.91 (br q, $J_{H,F}$ = 1.2, 3H, CH_3); $^{13}C\{^1H\}$ **NMR** (125.8 MHz, $CDCl_3$): δ = 166.1 (q, $J_{C,F}$ = 7.8 Hz, 1C, CCH_3), 140.4 (1C, C^3), 139.3 (1C, C^{7a}), 129.7 (1C, C^6), 127.2 (1C, C^{3a}), 124.9 (1C, C^5), 124.6 (q, $^1J_{C,F}$ = 257 Hz, 1C, CF_3), 121.1 (1C, C^4), 117.5 (1C, C^7), 18.3 (q, $J_{C,F}$ = 1.8 Hz, 1C, CH_3); ^{19}F **NMR** (376.5 MHz, $CDCl_3$): δ = −52.0 (br q, $J_{F,H}$ = 0.6 Hz, 3F, CF_3); **HRMS (EI)** calcd m/z for $C_{10}H_8F_3N_3$: 227.0665 [M$^+$], found: 227.0664 [M$^+$]; **Elemental Analysis** calcd (%) for $C_{10}H_8N_3F_3$: C 52.87, H 3.55, F 25.09, N 18.50, found: C 52.93, H 3.63, F 24.93, N 18.41; **CAS**: 1269630-44-5; **CCDC**: 792181.

(*E*)-*N*-(1-(3-(Adamantyl)-1*H*-pyrazol-1-yl)ethylidene)-1,1,1-trifluoromethanamine (55a) and (*E*)-*N*-(1-(5-(adamantyl)-1*H*-pyrazol-1-yl)ethylidene)-1,1,1-trifluoromethanamine (55b)

The title compounds were prepared from 1-(3-(adamantly)-1*H*-pyrazole (182 mg, 0.90 mmol, 1.5 eq.) according to the general procedure with an extended reaction time of 6 h. After extraction the regioisomers were separated and purified by flash

Experimental Part

chromatography (SiO$_2$, pentane/DCM 10:1 to 5:1) to yield major (79 mg, 42%) and minor regioisomere (5 mg, 3%) as colorless powder.

(55a). m.p.: 79-80 °C, R_f (pentane/CH$_2$Cl$_2$ 10:1): 0.56; **^{1}H NMR** (400.1 MHz, CDCl$_3$): δ = 8.32 (d, $J_{H,H}$ = 2.8 Hz, 1H, C^{5}H), 6.45 (d, $J_{H,H}$ = 2.8 Hz, 1H, C^{4}H), 2.80 (u q, $J_{H,F}$ = 1.6 Hz, 3H, CH$_3$), 2.10 (br s, 3H, CH$_{Ad}$), 2.00 (m 6H, 3x CH$_{2,Ad}$),1.83 (m, 6H, 3x CH$_{2,Ad}$); **^{13}C{^{1}H} NMR** (100.6 MHz, CD$_2$Cl$_2$): δ = 167.4 (1C, C^3), 165.3 (q, $J_{C,F}$ = 7.5 Hz, 1C, CCH$_3$), 128.9 (1C, C^5), 124.7 (q, $^1J_{C,F}$ = 258 Hz, 1C, CF$_3$), 107.9 (1C, C^4), 42.2 (3C, CH$_{2,Ad}$), 37.1 (3C, CH$_{2,Ad}$), 34.9 (1C, C$_{Ad}$), 29.0 (3C, CH$_{Ad}$), 17.0 (1C, CH$_3$); **^{19}F NMR** (376.5 MHz, CDCl$_3$): δ = –53.0 (u q, $J_{F,H}$ = 0.8 Hz, 3F, CF$_3$); **HRMS (EI)** calcd *m/z* for C$_{16}$H$_{20}$F$_3$N$_3$: 311.1604 [M$^+$], found: 311.1605 [M$^+$]; **Elemental Analysis** calcd (%) for C$_{16}$H$_{20}$F$_3$N$_3$: C 61.72, H 6.47, F 18.31, N 13.50, found: C 61.76, H 6.45, F 18.39, N 13.24.

(55b). R_f (pentane/CH$_2$Cl$_2$ 10:1): 0.22; **^{1}H NMR** (400.1 MHz, CD$_2$Cl$_2$): δ = 7.62 (d, $J_{H,H}$ = 1.6 Hz, 1H, C^{3}H), 6.35 (d, $J_{H,H}$ = 1.6 Hz, 1H, C^{4}H), 2.86 (u q, $J_{H,F}$ = 1.2 Hz, 3H, CH$_3$), 2.21 (m, 6H, 3x CH$_{2,Ad}$), 2.08 (br s, 3H, 3x CH$_{Ad}$), 1.79 (m, 6H, 3x CH$_{2,Ad}$); **^{13}C{^{1}H} NMR** (100.6 MHz, CD$_2$Cl$_2$): δ = 167.9 (u q, $J_{C,F}$ = 7.5 Hz, 1C, CCH$_3$), 158.0 (1C, C^5), 142.3 (1C, C^3), 124.4 (q, $^1J_{C,F}$ = 259 Hz, 1C, CF$_3$), 110.2 (1C, C^4), 40.6 (3C, 3x CH$_{2,Ad}$), 36.8 (3C, 3x CH$_{2,Ad}$), 36.2 (1C, C$_{Ad}$), 29.1 (3C, 3x CH$_{Ad}$), 20.9 (1C, CH$_3$); **^{19}F NMR** (376.5 MHz, CDCl$_3$): δ = –53.6 (s, 3F, CF$_3$); **HRMS (EI)** calcd *m/z* for C$_{16}$H$_{20}$F$_3$N$_3$: 311.1604 [M$^+$], found: 311.1608 [M$^+$].

(E)-N-(1-(3-(Adamantyl)-1H-pyrazol-1-yl)propylidene)-1,1,1-trifluoromethanamine (56)

The title compound was prepared from 1-(3-(adamantly)-1H-pyrazole (182 mg, 0.90 mmol, 1.5 eq.) according to the general procedure using EtCN (6 mL) instead of CH$_3$CN and with an extended reaction time of 6 h. After the reaction the solvent was removed under reduced pressure and the residue purified by flash chromatography (SiO$_2$, pentane) to yield titled compound as colorless powder (65 mg, 35%). **m.p:** 64 °C; R_f (pentane): 0.2; **^{1}H NMR** (400 MHz, CD$_2$Cl$_2$): δ = 8.27 (d, $J_{H,H}$ = 2.8 Hz, 1H, C^{5}H), 6.42 (d, $J_{H,H}$ = 2.8 Hz, 1H, C^{4}H), 3.20 (q, $J_{H,F}$ = 7.6 Hz, 2H, CH$_2$CH$_3$), 2.10 (br s, 3H, 3x CH$_{Ad}$), 1.99 (m, 6H, 3x CH$_{2,Ad}$), 1.79 (m, 6H, 3x CH$_{2,Ad}$), 1.34 (t, $J_{H,H}$ = 7.6 Hz, 3H, CH$_3$); **^{13}C{^{1}H} NMR** (75.5 MHz, CD$_2$Cl$_2$): δ = 169.1 (q, $J_{C,F}$ = 7.2 Hz, 1C, CCH$_2$CH$_3$), 167.0 (1C, C^3), 128.6 (1C, C^5), 124.3 (q, $^1J_{C,F}$ = 258 Hz, 1C, CF$_3$), 107.1 (1C, C^4),41.8 (3C, 3x CH$_{2,Ad}$), 36.7 (3C, 3x CH$_{2,Ad}$), 34.4 (1C, C$_{Ad}$), 28.6 (3C, 3x CH$_{Ad}$), 24.4 (1C, CH$_2$CH$_3$), 12.6 (1C, CH$_2$CH$_3$); **^{19}F NMR** (376.5 MHz, CDCl$_3$): δ = –53.0 (s, 3F, CF$_3$); **HRMS (EI)** calcd *m/z* for C$_{17}$H$_{22}$F$_3$N$_3$: 325.1761 [M$^+$], found: 325.1754 [M$^+$]; **Elemental Analysis** calcd (%) for $C_{17}H_{22}F_3N_3$: C 62.75, H 6.81, F 17.52, N 12.91, found: C 62.74, H 6.77, F 17.60, N 12.86.

(*E*)-*N*-(1-(3-(*tert*-Butyl)-1*H*-pyrazol-1-yl)ethylidene)-1,1,1-trifluoromethanamine (57)

The title compound was prepared from 3-*tert*-butyl-1*H*-pyrazole (112 mg, 0.90 mmol, 1.5 equiv) according to the general procedure, the reaction time was elongated to 6 h and purified by flash chromatography (SiO$_2$, pentane/CH$_2$Cl$_2$ 5:1) to yield corresponding imine as colorless oil containing sideproducts. R_f (pentane/CH$_2$Cl$_2$ 20:1): 0.45; **^{1}H NMR** (500.2 MHz, CDCl$_3$): δ = 8.29 (d, $J_{H,H}$ = 2.9 Hz, 1H, C^5*H*), 6.37 (d, $J_{H,H}$ = 2.9 Hz, 1H, C^4*H*), 2.78 (br q, $J_{H,F}$ = 1.4 Hz, 3H, N^{1}CC*H*$_3$), 1.31 (s, 9H, C*H*$_{3,tBu}$); **^{13}C{^{1}H} NMR** (125.8 MHz, CDCl$_3$): δ = 167.0 (1C, *C*^{3}), 164.6 (q, $J_{C,F}$ = 7.5 Hz, 1C N^{1}C*C*H$_3$), 129.0 (1C, *C*^{5}), 124.3 (q, $^1J_{C,F}$ = 258 Hz, 1C, *C*F$_3$), 108.1 (1C, *C*^{4}), 32.7 (1C, *C*(CH$_3$)$_3$), 30.0 (3C, C(*C*H$_3$)$_3$), 16.9 (br q, $J_{C,F}$ = 1.3 Hz, 1C, N^{1}C*C*H$_3$); **^{15}N NMR** (71 MHz, CDCl$_3$): δ = 237.8 (1N, *N*CF$_3$), 224.0 (1N, *N*^{1}), *N*^{2} not observed; **^{19}F NMR** (376.5 MHz, CDCl$_3$): δ = −52.8 (s, 3F, C*F*$_3$).

(*E*)-*N*-(1-(3-(Mesityl)-1*H*-pyrazol-1-yl)ethylidene)-1,1,1-trifluoromethanamine (58)

The title compound was prepared from 3-mesityl-1*H*-pyrazole (168 mg, 0.90 mmol, 1.5 equiv) according to the general procedure and purified by flash chromatography (SiO$_2$, pentane/CH$_2$Cl$_2$ 5:1) to yield corresponding imine (80.3 mg, 45%) as white crystals. **Mp:** 72 °C; R_f (pentane/CH$_2$Cl$_2$ 5:1): 0.3; **^{1}H NMR** (700.1 MHz, CDCl$_3$): δ = 8.49 (d, $J_{H,H}$ = 2.8 Hz, 1H, C^5*H*), 6.96 (s, 2H, 2x C*H*$_{m\text{-Mes}}$), 6.43 (d, $J_{H,H}$ = 2.8 Hz, 1H, C^4*H*), 2.82 (br q, $J_{H,F}$ = 2.8 Hz, 3H, N^{1}CC*H*$_3$), 2.33 (s, 3H, C*H*$_{3,p\text{-Mes}}$), 2.15 (s, 6H, 2x C*H*$_{3,o\text{-Mes}}$); **^{13}C{^{1}H} NMR** (176.0 MHz, CDCl$_3$): δ = 164.7 (q, $J_{C,F}$ = 7.0 Hz, 1C, N^{1}C*C*H$_3$), 156.1 (1C, *C*^{3}), 138.6 (1C, *C*$_{p\text{-Mes}}$), 137.3 (2C, 2C *C*$_{o\text{-Mes}}$), 129.3 (1C, *C*$_{ipso\text{-Mes}}$), 129.0 (1C, *C*^{5}), 128.6 (2C, *C*H$_{m\text{-Mes}}$), 124.1 (q, $^1J_{C,F}$ = 259 Hz, 1C, *C*F$_3$), 112.3 (1C, *C*^{4}), 21.2 (1C, *C*H$_{3,p\text{-Mes}}$), 20.6 (2C, 2x *C*H$_{3,o\text{-Mes}}$), 17.2 (br q, $J_{C,F}$ = 1.5 Hz, 1C, N^{1}C*C*H$_3$); **^{15}N NMR** (40.6 MHz, CDCl$_3$): δ = 241.7 (1N, *N*CF$_3$), 227.9 (1N, *N*^{1}), *N*^{2} not observed; **^{19}F NMR** (376.5 MHz, CDCl$_3$): δ = −53.0 (s, 3F, C*F*$_3$); **HRMS (EI)** calcd *m/z* for $C_{15}H_{16}F_3N_3$: 295.1291 [M$^+$], found: 295.1292 [M$^+$]; **Elemental Analysis** calcd (%) for $C_{15}H_{16}F_3N_3$: C 61.01, H 5.46, F 19.30, N 14.23; found: C 61.09, H 5.58, F 19.16, N 14.27; **CAS:** 1269630-39-8.

(*E*)-*N*-(1-(3,5-Di-*tert*-butyl-1*H*-pyrazol-1-yl)ethylidene)-1,1,1-trifluoromethanamine (59)

The title compound was prepared from 3,5-di-*tert*-butylpyrazole (162 mg, 0.90 mmol, 1.5 eq.) according to the general procedure with an extended reaction time of 16 h and purified by flash chromatography (pentane/benzene 100:1) to yield title compound (81 mg, 47%) as a colourless solid. **m.p.**: 37 °C; R_f (pentane/benzene 100:1): 0.55; **^{1}H NMR** (500.2 MHz, CDCl$_3$): δ = 6.17 (s, 1H, C^{4}H), 2.77 (br q, $J_{H,F}$ = 1.1 Hz, 3H, CH$_3$), 1.44 (s, 9H, C^{5}C(CH$_3$)$_3$), 1.28 (s, 9H,3x C^{3}C(CH$_3$)$_3$); **^{13}C{^{1}H} NMR** (125.8 MHz, CDCl$_3$): δ = 166.7 (q, $J_{C,F}$ = 7.5 Hz, 1C, N^{1}CCH$_3$), 163.3 (1C, C^3), 157.5 (1C, C^5), 124.3 (q, $^1J_{C,F}$ = 259 Hz, 1C, CF$_3$), 107.6 (1C, C^4), 33.7 (1C,C^{3}C(CH$_3$)$_3$), 32.5 (1C, C^{5}C(CH$_3$)$_3$), 30.0 (3C, 3x C^{5}C(CH$_3$)$_3$), 29.9 (3C,3x C^{3}C(CH$_3$)$_3$), 20.3 (q, $J_{C,F}$ = 1.8 Hz, 1C, CH$_3$); **^{15}N NMR** (40.6 MHz, CDCl$_3$): δ = 294.6 (1N, N^2), 241.6 (1N, NCF$_3$); 210.3 (1N, N^1); **^{19}F NMR** (376.5 MHz, CDCl$_3$): δ = −53.1 (br q, $J_{F,H}$ = 0.8 Hz, 3F, CF$_3$); **HRMS (EI)** calcd *m/z* for C$_{14}$H$_{22}$F$_3$N$_3$: 289.1761 [M$^+$], found: 289.1763 [M$^+$]; **Elemental Analysis** calcd (%) for C$_{14}$H$_{22}$N$_3$F$_3$: C 58.12, H 7.66, F 19.70, N 14.52, found: C 57.97, H 7.58, F 19.72, N 14.42; **CAS**: 1269630-40-1.

(*E*)-*N*-(1-(3,5-Diphenyl-1*H*-pyrazol-1-yl)ethylidene)-1,1,1-trifluoromethanamine (60)

The title compound was prepared from 3,5-diphenyl-1*H*-pyrazole (198 mg, 0.90 mmol, 1.5 equiv) according to the general procedure and was filtered, after extraction, through a silica pad with pentane/CH$_2$Cl$_2$ = 2:1. The brownish solid was triturated in a little pentane and the resulting white crystals were washed with a little cold pentane. A second crop of material was obtained by removing the solvent of the decanted solution under reduced pressure and trituration of that solid. The combined crystals were dried under vacuum to yield the title compound (92.2 mg, 47%) as white crystals. Single crystals for X-ray analysis were obtained by slow evaporation of pentane. **m.p.**: 108 °C; **^{1}H NMR** (500.2 MHz, CDCl$_3$): δ = 7.90 (d, $J_{H,H}$ = 7.2 Hz, 2H, C^3-CH$_{o\text{-Ph}}$), 7.47 (t, $J_{H,H}$ = 7.3 Hz, 2H, C^3-CH$_{m\text{-Ph}}$), 7.43-7.39 (*m*, 6H, CH$_{Ph}$), 6.78 (s, 1H, H^4), 2.90 (s, 3H, CH$_3$); **^{13}C{^{1}H} NMR** (125.8 MHz, CDCl$_3$): δ = 165.1 (q, $J_{C,F}$ = 7.2 Hz, 1C, N^{1}CCH$_3$), 153.7 (1C, C^3), 147.8 (1C, C^5), 131.9 (1C, C^{5}C$_{ipso\text{-Ph}}$), 131.8 (1C, C^{3}C$_{ipso\text{-Ph}}$), 129.4 (1C, CH$_{Ph}$), 129.02 (1C, CH$_{Ph}$), 128.98 (1C, CH$_{Ph}$), 128.6 (1C, CH$_{Ph}$), 128.0 (1C, CH$_{Ph}$), 126.3 (1C, C^{3}CH$_{o\text{-Ph}}$), 123.6 (q, $^1J_{C,F}$ = 259 Hz, 1C, CF$_3$), 110.8 (1C, C^4), 18.9 (q, $J_{C,F}$ = 1.9 Hz, 1C, CH$_3$); **^{15}N NMR** (40.6 MHz, CDCl$_3$): δ = 255.2 (1N, NCF$_3$), 220.2 (1N, N^1), N^2 not observed; **^{19}F NMR** (376.5 MHz, CDCl$_3$): δ = −54.0 (s, 3F, CF$_3$); **HRMS (EI)** calcd *m/z* for C$_{18}$H$_{14}$F$_3$N$_3$: 329.1135 [M$^+$], found: 329.1136 [M⁺]; **Elemental Analysis** calcd (%) for $C_{18}H_{14}N_3F_3$: C 65.65, H 4.28, F 17.31, N 12.76, found C 65.72, H 4.53, F 17.24, N 12.64; **CAS**: 1269630-41-2; **CCDC**: 792182.

(*E*)-*N*-(1-(4-Methyl-1*H*-pyrazol-1-yl)ethylidene)-1,1,1-trifluoromethanamine (61)

The title compound was prepared from 4-methyl-1*H*-pyrazole (72 μL, 0.90 mmol, 1.5 equiv) according to the general procedure and purified by flash chromatography (SiO₂, pentane/CH₂Cl₂ = 5:1) to yield corresponding imine (62 mg, 53%) as a colourless oil. R_f (pentane/CH₂Cl₂ 5:1): 0.26; **¹H NMR** (700.1 MHz, CDCl₃): δ = 8.16 (s, 1H, C⁵*H*), 7.57 (s, 1H, C³*H*), 2.75 (br q, $J_{H,F}$ = 1.2 Hz, 3H, C*H*₃), 2.11 (s, 3H, C⁴C*H*₃); **¹³C{¹H} NMR** (176.0 MHz, CDCl₃): δ = 164.2 (q, $J_{C,F}$ = 7.3 Hz, 1C, N¹*C*CH₃), 146.2 (1C, *C*³), 126.6 (1C, *C*⁵), 124.2 (q, $^1J_{C,F}$ = 259 Hz, *C*F₃), 121.4 (1C, *C*⁴), 16.7 (1C, *C*H₃), 9.0 (1C, *C*⁴CH₃); **¹⁹F NMR** (376.5 MHz, CDCl₃): δ = −53.0 (s, 3F, C*F*₃). **HRMS (EI)** calcd *m/z* for $C_7H_8F_3N_3$: 191.0665 [M⁺], found: 191.0662 [M⁺].

(*E*)-Ethyl 1-(1-(trifluoromethylimino)ethyl)-1*H*-pyrazole-4-carboxylate (62)

The title compound was prepared from ethyl 4-pyrazolecarboxylate (126 mg, 0.90 mmol, 1.5 equiv) according to the general procedure. Before extraction (methoxycarbonylsulfamoyl)-triethylammonium hydroxide (214 mg, 0.90 mmol, 1.5 eq.) dissolved in CH₂Cl₂ (0.6 mL) was added to the crude product mixture and stirred at 60 °C for an additional 30 min. The solvent was subsequently removed under reduced pressure and the residue was purified by flash chromatography (SiO₂, pentane/CH₂Cl₂ 2:1) and dried at 5 °C during 2 h under vacuum to yield title compound (56.4 mg, 38%) as a colourless oil. R_f 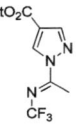 (pentane/CH₂Cl₂ 2:1): 0.23; **¹H NMR** (400.1 MHz, CDCl₃): δ = 8.85 (s, 1H, C⁵*H*), 8.07 (s, 1H, C³*H*), 4.32 (q, $J_{H,H}$ = 7.2 Hz, 2H, C*H*₂), 2.80 (br q, $J_{H,F}$ = 1.2 Hz, 3H, C*H*₃), 1.36 (t, $J_{H,H}$ = 7.1 Hz, 3H, CH₂C*H*₃); **¹³C{¹H} NMR** (125.8 MHz, CDCl₃): δ = 164.6 (q, $J_{C,F}$ = 7.3 Hz, 1C, N¹*C*CH₃), 162.1 (1C, *C*=O), 144.4 (1C, *C*³), 131.6 (1C, *C*⁵), 123.6 (q, $^1J_{C,F}$ = 261 Hz, 1C, *C*F₃), 119.2 (1C, *C*⁴), 61.1 (1C, *C*H₂), 17.1 (q, $J_{C,F}$ = 1.7 Hz, 1C, *C*H₃), 14.4 (1C, CH₂*C*H₃); **¹⁵N NMR** (40.6 MHz, CDCl₃): δ = 307.2 (*N*²), 250.6 (*N*CF₃), 230.7(*N*¹); **¹⁹F NMR** (376.5 MHz, CDCl₃): δ = −54.1 (s, 3F, C*F*₃); **HRMS (EI)** calcd *m/z* for $C_9H_{10}F_3N_3O_2$: 249.0720 [M⁺], found: 249.0718 [M⁺]; **Elemental Analysis** calcd (%) for $C_9H_{10}F_3N_3O_2$ C 43.38, H 4.04, F 22.87, N 16.86, O 12.84, found: C 43.32, H 4.05, F 22.70, N 16.67; **CAS**: 1269630-43-4.

Experimental Part

5.3.2 Direct *N*-Trifluoromethylation

N-(Diphenylmethylene)-1,1,1-trifluoromethanamine (64)

A Young Schlenk (20 mL) was charged with **1a** (198 mg, 0.60 mmol) and chlorotris(trimethylsilyl)silane (170 mg, 0.60 mmol, 1.0 equiv). Benzophenone imine (150 mL, 0.90 mmol, 1.5 equiv) dissolved in CH_3CN (6 mL) was added. The mixture was stirred at 60 °C for 18 h. Water was then added and the mixture was extracted with pentane (3x 15 mL). The combined organic phases were dried over Na_2SO_4 and the solvent of the removed under reduced pressure. The crude product was purified by flash chromatography (SiO_2, pentane/Et_2O gradient 200:1 to 50:1) to yield the titled compound as colorless oil with minor impurities (97 mg, 65%). The product decomposes over time in solution as well as in solid state even at -18 °C. R_f (pentane/Et_2O 50:1): 0.4; **^{1}H NMR** (500.2 MHz, $CDCl_3$): δ = 7.76 (d, $J_{H,H}$ = 7.4 Hz, 2H, $C_{Ar}H$), 7.58-7.47 (m, 4H, $C_{Ar}H$), 7.43 (t, $J_{H,H}$ = 7.8 Hz, 1H, $C_{Ar}H$), 7.32 (d, $J_{H,H}$ = 7.4 Hz, 1H, $C_{Ar}H$); **^{13}C{^{1}H} NMR** (125.8 MHz, $CDCl_3$): δ = 178.9 (br q, $J_{C,F}$ = 8.0 Hz, C_{imin}), 137.8 ($C_{q,Ar}$), 135.9 ($C_{q,Ar}$), 133.0 ($C_{Ar}H$), 130.4 ($C_{Ar}H$), 128.4 ($C_{Ar}H$); 128.0 ($C_{Ar}H$), 127.4 (q, $J_{C,F}$ = 1.7 Hz, $C_{Ar}H$), 123.7 (q, $^1J_{C,F}$ = 262 Hz, CF_3); **^{15}N NMR** (40.6 Hz, $CDCl_3$): δ = 252.4 (NCF_3); **^{19}F NMR** (376.5 MHz, $CDCl_3$): –51.9 (CF_3); **HRMS (EI)** calcd m/z for $C_{14}H_{10}F_3N$: 249.0765 [M$^+$], found: 249.0759 [M$^+$].

General Procedure for the synthesis of 4,5-disubstituted 1,2,3-triazoles

Caution: This reaction produces toxic and explosive hydrogen azide. Furthermore the reaction vessel is under overpressure. Therefore the reaction should be conducted exclusively behind a safety shield in a well ventilated laboratory hood. This general procedure is a modification of the Huisgen azide-alkyne dipolar cycloaddition (AAC).[132] In a closed Young Schlenk (10 mL), alkyne (5 mmol) and $TMSN_3$ were stirred without solvent for 12-120 h at 100-180 °C. The reaction vessel was then cooled to room temperature and the contents diluted with diethyl ether. The resulting solution was washed with deionized water, dried over $MgSO_4$ and the remaining solvents and any residual alkyne were removed under reduced pressure.

Dimethyl 2*H*-1,2,3-triazole-4,5-dicarboxylate (74)

Dimethyl acetylenedicarboxylate (0.62 mL, 5 mmol) and $TMSN_3$ (0.74 mL, 5.5 mmol, 1.1 equiv) were heated following the general procedure at 100 °C for 12 h. The crude product was recrystallized from benzene (30 mL) to yield **74** as a white crystalline material (0.35 g, 38%).

¹H NMR (700 MHz, CDCl₃): δ = 14.20 (br s, 1H, N*H*), 4.01 (s, 6H, CO₂C*H*₃); **¹³C{¹H} NMR** (176 MHz, CDCl₃): 160.5 (br s, *C*=O), 139.7 (br s, *C*⁴, *C*⁵), 53.4 (s, 2C, 2x *C*H₃); **HRMS (EI)** calcd *m/z* for C₆H₇N₃O₄: 185.0437 [M⁺], found: 185.0431 [M⁺, 5%], 154.0246 [M⁺-OCH₃, 100%]; **CAS**: 707-94-8.

4,5-Diphenyl-1*H*-1,2,3-triazole (75)

1,2-Diphenylethyne (891 mg, 5 mmol) and TMSN₃ (0.74 mL, 5.5 mmol, 1.1 equiv) were heated according to the general procedure at 180 °C for 5 d. The crude product was diluted in MeOH and stirred for 1 h at reflux. The solvent was removed and **75** was recovered as a light-brownish crystalline material (0.95 g, 69%). **¹H NMR** (700 MHz, CDCl₃): δ = 12.80 (br s, 1H, N*H*), 7.58-7.56 (m, 4H, 4x C*H*₀₋Ph), 7.38-7.36 (m, 6H, 4x C*H*ₘ₋Ph, 2x C*H*ₚ₋Ph); **¹³C{¹H} NMR** (176 MHz, CDCl₃): δ = 142.9 (*C*⁴, *C*⁵), 130.2 (2x *C*ᵢₚₛₒ₋Ph), 128.7 (2x *C*ₚ₋Ph), 128.6 (4x *C*ₘ₋Ph), 128.2 (4x *C*₀₋Ph); **HRMS (EI)** calcd *m/z* for C₁₄H₁₁N₃: 221.0952 [M⁺], found: 221.0948 [M⁺]; **CAS**: 5533-73-3.

General procedure for N-trifluoromethylation of nitrogen-heterocycles

A flame-dried two-necked flask (25 mL) with reflux condenser was charged with silica sulphuric acid (SSA, 2.8 mg) and azole (0.55 mmol, 1.1 equiv), HMDS (5.5 mL) was added and the mixture was heated to reflux for 2 h. To remove SSA, the mixture was cooled to 100 °C and the solution was filtered off into an Young Schlenk (20 mL) via a filter canula. The original reaction vessel was rinsed with toluene (3x 0.5 mL). After cooling to room temperature, all volatile compounds were removed under reduced pressure (15 mbar, 30 min 10⁻³ mbar). In a glovebox, the residue was redissolved in CH₂Cl₂ (0.33 mL) and LiNTf₂ (2.9 mg, 0.01 mmol, 2 mol-%.) was added. After shaking, **1a** (165 mmol, 0.5 mmol, 1 equiv) and subsequently HNTf₂ (16.9 mg, 0.06 mmol, 12 mol-%) were added, and the neck of the vessel was rinsed with CH₂Cl₂ (50 μL). The resulting clear solution was then stirred at 35 °C (bath temperature) in a closed Young Schlenk for 15 h. The solvent was then removed at reduced pressure (650 mbar, 40 °C).

1-(Trifluoromethyl)-1*H*-benzo[*d*][1,2,3]triazole (40a)

1*H*-4-Benzo[*d*][1,2,3]triazole (65 mg, 0.55 mmol, 1.1 equiv) was reacted with **1a** according to the general procedure. To facilitate purification, 3HF-NEt₃ (33 μL, 0.20 mmol, 0.4 equiv) was added after the reaction was completed. After 30 min stirring at room temperature the solvent was evaporated. The residue was purified by flash chromatography (Alox N, pentane/DCM 50:1) to yield (**40a**) as a colorless liquid that shows identical spectra to those reported in literature (59.9 mg, 64%).[109]

1-(Trifluoromethyl)-1H-benzo[d][1,2,3]triazole (40a) and 2-(trifluoromethyl)-2H-benzo[d][1,2,3]triazole (40b)

1H-4-Benzo[d][1,2,3]triazole (65 mg, 0.55 mmol, 1.1 equiv) was reacted with **1a** according to the general procedure with omission of the LiNTf$_2$ and by using BF$_3$-OEt$_2$ (3.1 µl, 0.025 mmol, 5 mol-%) as acid instead of HNTf$_2$. To facilitate purification, 3HF-NEt$_3$ (33 µL, 0.20 mmol, 0.4 equiv) was added after the reaction was completed, and after 30 min stirring at room temperature, the reaction mixture was evaporated. The isomeric mixture was separated by flash chromatography (Alox B, act. I, pentane/Et$_2$O gradient, 200:1 to 2:1) to yield **40a** (41.2 mg, 44%) and after further purification by bulb-to-bulb distillation (0 °C to −78 °C) isomere **40b** (22.5 mg, 24%). **4a** shows identical spectra to those reported in literature.[109]

(**40a**). ^{1}**H NMR** (700.1 MHz, CDCl$_3$): δ = 8.16 (Ψdt, $J_{H,H}$ = 8.3 Hz, $J_{H,H}$ = 0.9 Hz, 1H, C^{4}H), 7.76 (Ψdq, J = 8.3 Hz, J = 1.1 Hz, 1H, C^{7}H), 7.71 (ddd, $J_{H,H}$ = 8.2 Hz, $J_{H,H}$ = 6.9 Hz, $J_{H,H}$ = 1.0 Hz, 1H, C^{6}H), 7.56 (ddd, $J_{H,H}$ = 8.1 Hz, $J_{H,H}$ = 7.0 Hz, $J_{H,H}$ = 1.1 Hz, 1H, C^{5}H); ^{13}**C{^{1}H} NMR** (75.5 MHz, CDCl$_3$): δ = 146.0 (q, $J_{C,F}$ = 0.8 Hz, 1C, C^{3a}), 130.7 (1C, C^{7a}), 130.4 (1C, C^{6}H), 125.9 (1C, C^{5}H), 121.0 (1C, C^{4}H), 119.2 (q, $J_{C,F}$ = 266 Hz, 1C, CF$_3$), 110.4 (q, $J_{C,F}$ = 2.3 Hz, 1C, C^{7}H); ^{19}**F NMR** (659 MHz, CDCl$_3$): δ = −57.9; **HRMS (EI)** calcd m/z for C$_7$H$_4$N$_3$F$_3$: 187.0352 [M$^+$]; found: 187.0356 [M$^+$]; **Elemental Analysis** calcd (%) for C$_7$H$_4$F$_3$N$_3$: C 44.93, H 2.15, N 22.46, F 30.46; found: C 40.7, H 2.24, N 22.42; **CAS**: 328406-11-7.

(**40b**). ^{1}**H NMR** (500 MHz, CDCl$_3$): δ = 8.00-7.90 (m, 2H), 7.59-7.47 (m, 2H); ^{13}**C{^{1}H} NMR** (126 MHz, CD$_2$Cl$_2$): δ = 145.2 (C^{3a}, C^{7a}), 129.7 (C^4, C^7), 119.2 (C^5, C^6); CF$_3$ not observed; ^{15}**N NMR** (40.6 MHz, CD$_2$Cl$_2$): δ = 315 (N^1, N^3); 256 (d, $J_{F,N}$ = 19 Hz, N^2); ^{19}**F NMR** (659 MHz, CDCl$_3$): δ = −61.2; **HRMS (EI)** calcd m/z for C$_7$H$_4$N$_3$F$_3$: 187.0352 [M$^+$], found: 187.0352 [M$^+$]; **Elemental Analysis** calcd (%) for C$_7$H$_4$F$_3$N$_3$: C 44.93, H 2.15, N 22.46, F 30.46; found: C 44.77, H 2.39, N 22.17, F 30.20.

3-(1-Adamantyl)-1-(trifluoromethyl)-1H-pyrazole (79)

3-(1-Adamantyl)-1H-pyrazole (111 mg, 0.55 mmol, 1.1 equiv) was reacted with **1a** according to the general procedure and the residue was purified by flash chromatography (SiO$_2$, pentane/CH$_2$Cl$_2$ 50:1) to yield the title compound

Experimental Part

as a colorless powder (83.8 mg, 62%). Crystals for X-Ray analysis were obtained by sublimation (15 mbar, 25 °C). **m.p.**: 31 °C; R_f (pentane/CH_2Cl_2 50:1): 0.3 (stained with $KMnO_4$); **^{1}H NMR** (700 MHz, $CDCl_3$): δ = 7.72 (d, $J_{H,H}$ = 2.8 Hz, 1H, C^3H), 6.34 (d, $J_{H,H}$ = 2.8 Hz, 1H, C^4H), 2.10 (br s, 3H, CH_{Ad}), 2.00 (d, $J_{H,H}$ = 2.8 Hz, 6H, $CH_{2,Ad}$), 1.80 (m, 6H, $CH_{2,Ad}$); **$^{13}C\{^1H\}$ NMR** (176 MHz, $CDCl_3$): δ = 166.3 (uq, $J_{C,F}$ = 0.5 Hz, C^3), 128.3 (C^5), 118.2 (q, $^1J_{C,F}$ = 261 Hz, CF_3), 104.9 (C^4), 42.1 (3x $CH_{2,Ad}$), 36.7 (3x $CH_{2,Ad}$), 34.3 (C_{Ad}), 28.4 (3x CH_{Ad}); **^{15}N NMR** (71 MHz, $CDCl_3$): δ = 200 ($J_{F,N}$ = 18 Hz, N^1), 172 (N^2); **^{19}F NMR** (659 MHz, $CDCl_3$): δ = –60.1; **HRMS (ESI)** calcd m/z for $C_{14}H_{18}F_3N_2$: 271.1417 [M$^+$]; found: 271.1409 [M$^+$]; **Elemental Analysis** calcd (%) for $C_{14}H_{17}F_3N_2$: C 62.21, H 6.34, F 21.09, N 10.36; found: C 62.42, H 6.51, F 20.81, N 10.10; **IR**(ATR): $\tilde{\nu}$ (cm^{-1})= 3146 (w), 3125 (w), 2910 (m), 2849 (m), 2676(m), 1535 (m), 1445 (w), 1412 (s), 1377 (s), 1341 (s), 1244 (s), 1169 (s), 1108 (s), 933 (s), 760 (m); **CCDC**: 841860.

1-(Trifluoromethyl)-3-(2,4,6-trimethylphenyl)-1H-pyrazole (80a) and 1-trifluoromethyl-5-(2,4,6-trimethylphenyl)pyrazole (80b)

3-(2,4,6-Trimethylphenyl)-1H-pyrazole (103 mg, 0.55 mmol, 1.1 equiv) was reacted with **1a** according to the general procedure using 14 mol-% HNTf$_2$ (19.2 mg) instead of 12 mol-%. The resulting regioisomeric mixture was separated by flash chromatography (Florisil, pentane/CH_2Cl_2 10:1) to yield pure **80a** (38.1 mg, 30%) and **80b** (15.9 mg, 25%, containing 3% **80a**). Single crystals for X-ray analysis of **80b** were obtained by sublimation (15 mbar, 60 °C), whereby the **70b** was obtained as single regioisomer in 15% yield.

(80a). **b.p.** (dec.): 173 °C; **^{1}H NMR** (700 MHz, $CDCl_3$): δ = 7.92 (d, $J_{H,H}$ = 2.5 Hz, 1H, C^3H), 6.96 (s, 2H, CH_{Mes}), 6.42 (d, $J_{H,H}$ = 2.5 Hz, 1H, C^4H), 2.35 (s, 3H, $CH_{3,p-Mes}$), 2.15 (s, 6H, 2x $CH_{3,o-Mes}$); **$^{13}C\{^1H\}$ NMR** (176 MHz, $CDCl_3$): δ = 155.6 (q, $J_{C,F}$ = 0.7 Hz, C^3), 138.7 (C_{p-Mes}), 137.7 (C_{o-Mes}), 129.3 ($C_{ipso-Mes}$), 128.9 (C^5), 128.7 (CH_{m-Mes}), 118.6 (q, $^1J_{C,F}$ = 263 Hz, CF_3), 110.7 (q, $J_{C,F}$ = 1 Hz, C^4), 21.5 ($CH_{3,p-Mes}$), 20.7 (2x $CH_{3,o-Mes}$); **^{15}N NMR** (71 MHz, $CDCl_3$): δ = 204 ($J_{F,N}$ = 17 Hz, N^1), 179 (N^2); **^{19}F NMR** (659 MHz, $CDCl_3$): δ = -60.7; **HRMS (EI)** calcd m/z for $C_{13}H_{13}F_3N_2$: 254.1026[M$^+$], found: 254.1026 [M$^+$, 100%], 239.0788 [M$^+$-CH$_3$, 16%], 170.0963 [$C_{12}H_{12}N^+$, 24%], 158.0971 [$C_{12}H_{14}\cdot^+$, 32%]; **Elemental Analysis** calcd (%) for $C_{13}H_{13}F_3N_2$: C 61.41, H 5.15, N 11.02, F 22.42; found: C 61.61, H 5.27, N 10.89, F 22.21; **IR**(ATR): $\tilde{\nu}$ (cm^{-1})= 3124 (w), 2955 (w), 2924 (w), 1728 (w), 1614 (w), 1541 (m), 1491 (w), 1422 (s), 1398 (s), 1292 (s), 1258 (s), 1167 (s), 1106 (s), 1080 (m), 1043 (m), 962 (s), 942 (s), 927 (s), 851 (m), 763 (s), 743 (m), 644 (m).

(80b). m.p.: 57 °C; ^{1}**H NMR** (700 MHz, CDCl$_3$): δ = 7.82 (d, $J_{H,H}$ = 1.4 Hz, 1H, C^{3}H), 6.97 (d, $J_{H,H}$ = 0.7 Hz, 2H, CH$_{Mes}$), 6.29 (d, $J_{H,H}$ = 1.4 Hz, 1H, C^{4}H), 2.36 (s, 3H, CH$_{3,p\text{-Mes}}$), 2.05 (s, 6H, 2x CH$_{3,o\text{-Mes}}$); ^{13}**C{^{1}H}-NMR** (176 MHz, CDCl$_3$): δ = 142.69 (C^5), 142.67 (C^3), 139.4 (C$_{p\text{-Mes}}$), 137.8 (C$_{o\text{-Mes}}$), 128.1 (C$_{m\text{-Mes}}$), 125.3 (C$_{ipso\text{-Mes}}$), 118.6 (q, $^1J_{C,F}$ = 265 Hz, CF$_3$), 110.2 (q, $J_{C,F}$ = 1.4 Hz, C^4), 21.1 (CH$_{3,p\text{-Mes}}$), 19.8 (2x CH$_{3,o\text{-Mes}}$); ^{15}**N NMR** (71 MHz, CDCl$_3$): δ = 205 ($J_{F,N}$ = 17 Hz, N^1), 180 (N^2); ^{19}**F NMR** (659 MHz, CDCl$_3$): δ = −58.0; **HRMS (EI)** calcd m/z for C$_{13}$H$_{13}$F$_3$N$_2$: 254.1026 [M$^+$], found: 254.1027 [M$^+$]; **Elemental Analysis** calcd (%) for C$_{13}$H$_{13}$F$_3$N$_2$: C 61.41, H 5.15, N 11.02, F 22.42; found: C 61.51, H 5.18, N 10.99, F 22.41; **IR(ATR)**: $\tilde{\nu}$ (cm^{-1})= 3138 (w), 3103 (w), 2947 (w), 2924 (w), 1612 (m), 1564 (w), 1451 (m), 1360 (s), 1285 (s), 1239 (s), 1224 (s), 1167 (m), 1094 (s), 1040 (m), 969 (m), 915 (s), 815 (s), 817 (m), 768 (m); **CCDC**: 841859.

Ethyl 3-methyl-1-(trifluoromethyl)-1H-pyrazole-4-carboxylate (81a)

Ethyl 3-methyl-1H-pyrazole-4-carboxylate (85 mg, 0.55 mmol, 1.1 equiv) was reacted with **1a** according to the general procedure and the resulting isomeric mixture was separated by flash chromatography (Florisil, pentane/Et$_2$O 15:1) to obtain the title compound **81a** (36.7 mg, 33%, containing 5% ethyl 5-methyl-1-(trifluoromethyl)-1H-pyrazole-4-carboxylate (**81b**)). Analytical samples and crystals of **81a** suitable for X-ray analysis were obtained by sublimation (15 mbar, 60 °C).

(81a). m.p.: 56 °C; ^{1}**H NMR** (700 MHz, CDCl$_3$): δ = 8.28 (s, 1H, CH), 4.35 (q, $J_{H,H}$ = 7 Hz, 2H, CH$_2$), 2.55 (s, 3H, C^{3}CH$_3$), 1.39 (t, $J_{H,H}$ = 7 Hz, 3H, CH$_2$CH$_3$); ^{13}**C{^{1}H} NMR** (176 MHz, CDCl$_3$): δ = 162.3 (C=O), 154.5 (C^4), 132.6 (C^5), 117.6 (q, $^1J_{C,F}$ = 264 Hz, CF$_3$), 115.6 (C^3), 60.7 (CH$_2$), 14.3 (CH$_2$CH$_3$), 13.6 (C^{3}CH$_3$); ^{15}**N NMR** (71 MHz, CDCl$_3$): δ = 202 ($J_{F,N}$ = 17 Hz, N^1), 177 (N^2); ^{19}**F NMR** (659 MHz, CDCl$_3$): δ = −61.0; **HRMS (ESI)** calcd m/z for C$_8$H$_9$F$_3$N$_2$O$_2$: 223.0689 [MH$^+$], found: 223.0685 [MH$^+$]; **Elemental Analysis** calcd (%) for C$_8$H$_9$F$_3$N$_2$O$_2$: C 43.25, H 4.08, N 12.65, F 25.65; found: C 43.22, H 4.14, N 12.65, F 25.37; **IR(ATR)**: $\tilde{\nu}$ (cm^{-1}) = 3133 (w), 3100 (w), 2989 (w), 2942 (w), 2914 (w), 2882 (w), 1727 (s), 1562 (m), 1491 (m), 1477 (m), 1429 (m), 1377 (m), 1305 (m), 1260 (s), 1181 (s), 1105 (s), 1086 (s), 1047 (m), 1021 (m), 1004 (m), 949 (s), 867 (m), 840 (m), 771 (s), 701 (s), 625 (s); **CCDC**: 841861.

(81b). ^{1}**H NMR** (700 MHz, CDCl$_3$): δ = 8.01 (s, 1H, CH), 4.35 (q, $J_{H,H}$ = 7 Hz, 2H, CH$_2$), 2.77 (u q, $J_{H,F}$ = 1.5 Hz, 3H, C^{5}CH$_3$), 1.39 (t, $J_{H,H}$ = 7 Hz, 3H, CH$_2$CH$_3$); ^{13}**C{^{1}H}-NMR** (176 MHz, CDCl$_3$): δ = 162.6 (C=O), 145.1 (C^4), 143.1 (C^3), 118.5 (q, $^1J_{C,F}$ = 265 Hz, CF$_3$), 115.4 (q, $J_{C,F}$ = 1.4 Hz, C^5), 60.6 (CH$_2$), 14.3 (CH$_2$CH$_3$), 10.5 (q, $J_{C,F}$ = 2.5 Hz, C^{5}CH$_3$); ^{15}**N NMR** (71 MHz, CDCl$_3$): δ =

Experimental Part

210 ($J_{F,N}$ = 18 Hz, N^1), 179 (N^2); **^{19}F NMR** (659 MHz, CDCl$_3$): δ = −57.4 (u q, $J_{F,H}$ = 1.5 Hz).

Ethyl 1-(trifluoromethyl)-1*H*-pyrazole-4-carboxylate (82)

Ethyl 1*H*-pyrazole-4-carboxylate (77 mg, 0.55 mmol, 1.1 equiv) was reacted with **1a** according to the general procedure using 14 mol-% HNTf$_2$ (19.7 mg) instead of 12 mol-%. The product was purified by flash chromatography (SiO$_2$, pentane/Et$_2$O 20:1) to yield the title compound as a colorless oil (13.5 mg, <13%) with contamination by reaction byproducts, as identified by NMR by comparison to literature data.[133] **CAS:** 29819-42-7.

4-Benzyl-1-(trifluoromethyl)-1*H*-pyrazole (83)

4-Benzyl-1*H*-pyrazole (87 mg, 0.55 mmol, 1.1 equiv) was reacted with **1a** according to the general procedure and purified by flash chromatography (SiO$_2$, pentane/Et$_2$O 50:1) to yield title compound as a colorless oil (74.5 mg, 66%). **R$_f$** (pentane/Et$_2$O 50:1): 0.3; **^{1}H NMR** (700 MHz, CD$_2$Cl$_2$): δ = 7.65 (s, 1H, C^{3}H), 7.61 (s, 1H, C^{5}H), 7.36 (Ψt, $J_{H,H}$ = 7.7 Hz, $J_{H,H}$ = 7.3 Hz, 2H, C$H_{m\text{-Ph}}$), 7.28 (Ψt, $J_{H,H}$ = 8.6 Hz, $J_{H,H}$ = 7.6 Hz, 1H, C$H_{p\text{-Ph}}$), 7.26 (d, $J_{H,H}$ = 7.5 Hz, 2H, C$H_{o\text{-Ph}}$), 3.90 (s, 2H, CH_2); **^{13}C{^{1}H} NMR** (176 MHz, CD$_2$Cl$_2$): δ = 144.0 (q, $J_{C,F}$ = 0.8 Hz, C^3), 139.7 (br s, $C_{ipso\text{-Ph}}$), 128.6 (2x C$H_{m\text{-Ph}}$), 128.5 (2x C$H_{o\text{-Ph}}$), 126.5 (C$H_{p\text{-Ph}}$), 126.3 (br s, C^5H), 124.0 (q, $J_{C,F}$ = 0.8 Hz, C^4), 118.1 (q, $^1J_{C,F}$ = 262 Hz, CF$_3$), 30.1 (br s, CH$_2$); **^{15}N NMR** (71 MHz, CDCl$_3$): δ = 204 ($J_{F,N}$ = 18 Hz, N^1), 182 (N^2); **^{19}F NMR** (659 MHz, CD$_2$Cl$_2$): δ = −61.6; **HRMS (EI)** calcd *m/z* for C$_{11}$H$_9$F$_3$N$_2$: 225.0634 [M-H$^+$], found 225.0634 [M-H$^+$]; **Elemental Analysis** calcd (%) for C$_{11}$H$_9$F$_3$N$_2$: C 58.41, H 4.01, N 12.38, F 25.20; found: C 58.48, H 3.99, N 12.44, F 25.07; **IR(ATR):** $\tilde{\nu}$ (cm^{-1}) = 3119 (w), 3089 (w), 3065 (w), 3030 (w), 2920(w), 1604 (w), 1582 (w), 1496 (s), 1418 (s), 1393 (m), 1295 (m), 12.18 (s), 1166 (s), 1075 (s), 1008 (m), 940 (s), 874 (w), 795 (w), 758 (m), 705 (m), 636 (m).

3,5-Dimethyl-1-(trifluoromethyl)-1*H*-pyrazole (84)

In a glovebox, 3,5-dimethyl-1-(trimethylsilyl)-1*H*-pyrazole (92 mg, 0.55 mol, 1.1 equiv) was dissolved in CH$_2$Cl$_2$ (0.33 mL) in a Young-Schlenk (10 mL) and LiNTf$_2$ (2.9 mg, 0.01 mol, 2 mol-%) was added. After shaking, **1a** (165 mg, 0.50 mol) and subsequently HNTf$_2$ (16.9 mg, 0.06 mmol, 12 mol-%) were added. Before closing the Young-Schlenk, the neck was rinsed with additional CH$_2$Cl$_2$ (50 μL). The solution was stirred at 35 °C (bath temperature) for 15 h. Afterwards, the solvent was carefully removed under reduced pressure and after two sequential bulb-to-

Experimental Part

bulb distillations (–25 °C to –78 °C, 10^{-3} mbar) the title compound was obtained as a highly volatile colorless liquid (12.5 mg, 15%). **^{1}H NMR** (700 MHz, CDCl$_3$): δ = 6.00 (s, 1H, C^{4}H), 2.41 (m, C^{5}CH$_3$), 2.78 (s, C^{3}CH$_3$); **^{13}C{^{1}H} NMR** (176 MHz, CDCl$_3$): δ = 151.9 (C^3), 141.1 (C^5), 119.3 (q, $^1J_{C,F}$ = 262 Hz, CF$_3$), 109.9 (q, $J_{C,F}$ = 1.4 Hz, C^4), 14.5 (C^{3}CH$_3$), 11.6 (q, $J_{C,F}$ = 2.6 Hz, C^{5}CH$_3$); **^{15}N NMR** (71 MHz, CDCl$_3$): δ = 199 ($J_{F,N}$ = 17 Hz, N^1), 174 (N^2); **^{19}F NMR** (659 MHz, CDCl$_3$): δ = –57.5; **HRMS (EI)** calcd m/z for C$_5$H$_7$F$_3$N$_2$: 164.0556 [M$^+$], found 164.0555 [M$^+$]; **Elemental Analysis** calcd (%) for C$_5$H$_7$F$_3$N$_2$: C 43.91, H 4.30, N 17.07; found: C 43.80, H 4.55, N 17.02.

3-Methyl-1-(trifluoromethyl)-5-(2,4,6-trimethylphenyl)-1H-pyrazole (85a) and 5-methyl-1-(trifluoromethyl)-3-(2,4,6-trimethylphenyl)-1H-pyrazole (85b)

5-Methyl-3-(2,4,6-trimethylphenyl)-1H-pyrazole (110 mg, 0.55 mmol, 1.1 equiv) was reacted with **1a** according to the general procedure. To facilitate purification, 3HF-NEt$_3$ (33 µL, 0.20 mmol, 0.4 equiv) was added after the reaction was completed, and after 30 min further stirring at room temperature, saturated aqueous NaHCO$_3$ solution was added and the mixture was extracted (3x pentane). The combined organic phases were washed with brine, dried over MgSO$_4$ and the solvent was removed under reduced pressure. The resulting regioisomeric mixture was separated by flash chromatography (Florisil, pentane/Et$_2$O 40:1) to yield **85b** as a colorless liquid (15.6 mg, 12%) and **85a** as colorless powder (54.2 mg, 40%). Single crystals in X-ray quality of **85a** were obtained by sublimation (15 mbar, 80 °C).

(**85a**). m.p. 113 °C; **^{1}H NMR** (500 MHz, CD$_2$Cl$_2$): δ = 6.96 (s, 2H, 2x CH$_{Mes}$), 6.10 (s, 1H, C^{4}H), 2.40 (s, 3H, C^{3}CH$_3$), 2.35 (s, 3H, CH$_{3,p-Mes}$), 2.10 (s, 6H, 2x CH$_{3,o-Mes}$); **^{13}C{^{1}H} NMR** (126 MHz, CD$_2$Cl$_2$): δ = 152.8 (C^3), 143.8 (C^5), 139.6 (C$_{p-Mes}$), 138.1 (2x C$_{o-Mes}$), 128.4 (CH$_{Mes}$), 126.0 (C$_{ipso-Mes}$), 119.1 (q, $^1J_{C,F}$ = 265 Hz, CF$_3$), 111.0 (q, $J_{C,F}$ = 1 Hz, C^4), 21.5 (CH$_{3,p-Mes}$), 20.2 (2x CH$_{3,o-Mes}$), 14.2 (C^{3}CH$_3$); **^{15}N NMR** (71 MHz, CDCl$_3$): δ = 200 ($J_{N,F}$ = 17 Hz, N^1), 174 (N^2); **^{19}F NMR** (659 MHz, CDCl$_3$): δ = –58.0; **HRMS (EI)** calcd m/z for C$_{14}$H$_{15}$F$_3$N$_2$: 268.1182 [M$^+$]; found: 268.1183 [M$^+$]; **Elemental Analysis** calcd (%) for C$_{14}$H$_{15}$F$_3$N$_2$: C 62.68, H 5.64, N 10.44, F 21.24; found: C 62.76, H 5.66, N 10.42, F 21.23; **IR(ATR)**: $\tilde{\nu}$ (cm^{-1}) = 3104 (w), 2983 (w), 2955 (w), 2924 (w), 2861 (w), 1614 (w), 1574 (w), 1498 (w), 1458 (w), 1408 (m), 1371 (m), 1342 (s), 1290 (m), 1236 (s), 1195 (s), 1151 (s), 1088 (s), 10.56 (m), 962 (m), 937 (m), 863 (m), 827 (m), 708 (m), 621 (m); **CCDC**: 841862.

(**85b**). **^{1}H NMR** (400 MHz, CDCl$_3$): δ = 6.96 (s, 2H, 2x CH$_{Mes}$), 6.15 (s, 1H, C^{4}H), 2.53 (s, 3H, C^{5}CH$_3$), 2.34 (s, 3H, CH$_{3,p-Mes}$), 2.14 (s, 6H, 2x CH$_{3,o-Mes}$); **^{13}C{^{1}H} NMR** (101 MHz, CDCl$_3$): δ = 154.0 (C^5), 141.4 (C^3), 138.4 (C$_{p-Mes}$), 137.5 (2x C$_{o-Mes}$), 129.7 (C$_{ipso-Mes}$), 128.5 (CH$_{Mes}$), 119.7 (q, $^1J_{C,F}$ = 265

Hz, CF$_3$), 111.3 (q, $J_{C,F}$ = 1.7 Hz, C^4), 21.2 (CH$_{3,p\text{-Mes}}$), 20.4 (2x CH$_{3,o\text{-Mes}}$), 11.9 (q, $J_{C,F}$ = 2.7 Hz, C^{5}CH$_3$); **^{15}N NMR** (71 MHz, CDCl$_3$): δ = 202 ($J_{F,N}$ = 18 Hz, N^1), 179 (N^2); **^{19}F NMR** (659 MHz, CDCl$_3$): δ = −57.4; **HRMS (EI)** calcd m/z for C$_{14}$H$_{15}$F$_3$N$_2$: 268.1182 [M$^+$], found: 268.1181 [M$^+$]; **Elemental Analysis** calcd (%) for C$_{14}$H$_{15}$F$_3$N$_2$: C 62.68, H 5.64, N 10.44, F 21.24; found: C 62.45, H 5.64, N 10.49.

2-(Trifluoromethyl)-2H-4,5,6,7-tetrahydroindazole (86) and 1-(trifluoromethyl)-1H-4,5,6,7-tetrahydroindazole (86)

1H-4,5,6,7-Tetrahydroindazole (69 mg, 0.55 mmol, 1.1 equiv) was reacted with **1a** according to the general procedure. To facilitate purification, 3HF-NEt$_3$ (33 µL, 0.20 mmol, 0.4 equiv) was added after the reaction was completed, and after 30 min stirring at room temperature, saturated aqueous NaHCO$_3$ was added and the mixture was extracted (3x pentane). The combined organic phases were washed with brine, dried over MgSO$_4$ and the solvent was removed under reduced pressure. The resulting regioisomeric mixture was separated by flash chromatography (Florisil, pentane/Et$_2$O 30:1, stained with CAM) to yield pure **86a** (28.4 mg, 30%) and **86b** (10.9 mg, 12%, containing ≤ 5% **86a**).

(**86a**). **^{1}H NMR** (700 MHz, CDCl$_3$): δ = 7.49 (s, 1H, CH), 2.75 (Ψt, $J_{H,H}$ = 6.6 Hz, $J_{H,H}$ = 6.3 Hz, 2H, C^{7}H$_2$), 2.60 (Ψt, $J_{H,H}$ = 6.3 Hz, $J_{H,H}$ = 6.4 Hz, C^{4}H$_2$), 1.88-1.84 (m, 2H, C^{6}H$_2$), 1.80-1.76 (m, 2H, C^{5}H$_2$); **^{13}C{^{1}H} NMR** (176 MHz, CDCl$_3$): δ = 154.4 (q, $J_{C,F}$ = 1 Hz, C^{7a}), 124.7 (C^3), 119.3 (q, $J_{C,F}$ = 1 Hz, C^{3a}), 118.4 (q, $J_{C,F}$ = 261 Hz, CF$_3$), 23.4 (C^7), 22.94 (C^6), 22.93 (C^5), 20.3 (C^4); **^{15}N NMR** (70.9 MHz, CDCl$_3$): δ = 198 ($J_{F,N}$ = 18 Hz, N^2), 170 (N^1); **^{19}F NMR** (659 MHz, CDCl$_3$): δ = −60.1; **HRMS (EI)** calcd m/z for C$_8$H$_9$F$_3$N$_2$: 190.0713 [M$^+$], found: 190.0713 [M$^+$]; **Elemental Analysis** calcd (%) for C$_8$H$_9$F$_3$N$_2$: C 50.53, H 4.77, N 14.73, F 29.97; found: C 50.52, H 4.74, N 14.95, F 29.75; **IR(ATR)**: $\tilde{\nu}$ (cm^{-1}) = 3107 (w), 2940 (m), 2862 (w), 1584 (w), 1479 (m), 1419 (s), 1370 (s), 1326 (w), 1293 (s), 1159 (s), 1078 (m), 941 (s), 849 (w), 824 (w), 787 (m), 707 (m), 634 (m).

(**86b**). **GC: 16b/16a** ≤ 95%; **^{1}H NMR** (700 MHz, CD$_2$Cl$_2$): δ = 7.50 (1H, CH), 2.78 (Ψt, $J_{H,H}$ = 6.3 Hz, $J_{H,H}$ = 6.2 Hz, 2H, C^{7}H$_2$), 2.53 (Ψt, $J_{H,H}$ = 6.2 Hz, $J_{H,H}$ = 6.0 Hz, 2H, C^{4}H$_2$), 1.91-1.85 (m, 2H, C^{6}H$_2$), 1.80-1.75 (m, 2H, C^{5}H$_2$); **^{13}C{^{1}H} NMR** (126 MHz, CD$_2$Cl$_2$): δ = 142.3 (q, $J_{C,F}$ = 0.5 Hz, C^{7a}), 140.1 (C^3), 120.2 (q, $J_{C,F}$ = 1.3 Hz, C^{3a}), 119.5 (q, $^1J_{C,F}$ = 260 Hz, CF$_3$), 22.7 (C^5), 22.7 (q, $J_{C,F}$ = 0.8 Hz, C^6), 22.3 (q, $J_{C,F}$ = 2.2 Hz, C^7), 20.7 (C^4); **^{15}N NMR** (70.9 MHz, CD$_2$Cl$_2$): δ = 200 ($J_{F,N}$ = 18 Hz, N^1), 175 (N^2); **^{19}F NMR** (659 MHz, CD$_2$Cl$_2$): δ = −58.4; **HRMS (EI)** calcd m/z for C$_8$H$_9$F$_3$N$_2$: 190.0713 [M$^+$], found: 190.0715 [M$^+$]; **Elemental Analysis** calcd (%)

for $C_8H_9F_3N_2$: C 50.53, H 4.77, N 14.73, F 29.97; found: C 50.40, H 4.69, N 14.92, F 30.16.

2-(Trifluoromethyl)-2H-indazole (87a)

1H-Indazole (66 mg, 0.55 mmol, 1.1 equiv) was reacted with **1a** according to the general procedure and the residue purified by flash chromatography (SiO$_2$, pentane/Et$_2$O 30:1) to yield 2-(trifluoromethyl)-2H-indazole (**87a**) as an colorless oil (35 mg, 39%, containing ≤ 2% **87b**). **b.p.** 190-194 °C; R_f (pentane/Et$_2$O 30:1): 0.4; **^{1}H NMR** (700 MHz, CDCl$_3$): δ = 8.39 (br s, C^3H), 7.79 (dq, $J_{H,H}$ = 8.9 Hz, $J_{H,H}$ = 1 Hz, 1H, C^4H), 7.71 (dt, $J_{H,H}$ = 8.7 Hz, $J_{H,H}$ = 1 Hz, 1H, C^7H), 7.42 (ddd, $J_{H,H}$ = 8.9 Hz, $J_{H,H}$ = 6.6 Hz, $J_{H,H}$ = 1 Hz, 1H, C^6H), 7.20 (ddd, $J_{H,H}$ = 8.7 Hz, $J_{H,H}$ = 6.6 Hz, $J_{H,H}$ = 1 Hz, 1H, C^5H); **^{13}C{^{1}H} NMR** (176 MHz, CDCl$_3$): δ = 150.5 (q, $J_{C,F}$ = 1Hz, C^{7a}), 128.9 (C^6), 124.3 (C^5), 121.6 (br s, C^3), 121.5 (q, $J_{C,F}$ = 0.8 Hz, C^{3a}), 120.8 (C^7), 118.7 (C^4), 118.7 (q, $^1J_{C,F}$ = 270 Hz, CF_3); **^{15}N NMR** (70.9 MHz, CDCl$_3$): δ = 216 ($J_{F,N}$ = 18 Hz, N^2), 160 (N^1); **^{19}F NMR** (659 MHz, CDCl$_3$): δ = –59.7; **HRMS (EI)** calcd m/z for $C_8H_5F_3N_2$: 186.0405 [M$^+$], found: 186.0404 [M$^+$]; **Elemental Analysis** calcd (%) for $C_8H_5F_3N_2$: C 51.62, H 2.71, N 15.05, F 30.62; found: C 51.35, H 2.72, N 15.10, F 30.53.

^{19}F NMR (568 MHz, CDCl$_3$): δ = –58.3 was tentatively assigned to 1-(trifluoromethyl)-1H-indazole (**87b**).

Dimethyl 1-(trifluoromethyl)-1H-1,2,3-triazole-4,5-dicarboxylate (88a)

Dimethyl 2H-1,2,3-triazole-4,5-dicarboxylate (102 mg, 0.55 mmol, 1.1 equiv) was reacted with **1a** according to the general procedure and the residue purified by bulb-to-bulb distillation (-12 °C to -78 °C at 10^{-3} mbar) to obtain the titled compound (**88a**) (30.4 mg, 24%, containing 8% dimethyl 2-(trifluoromethyl)-2H-1,2,3-triazole-4,5-dicacrobxylate (**88b**)). **^{1}H NMR** (700 MHz, CDCl$_3$): δ = 4.09 (s, 3H, $C^4CO_2CH_3$ major), 4.05 (s, 6H, 2x CO_2CH_3 minor), 4.03 (s, 3H, $C^5CO_2CH_3$ major); **^{13}C{^{1}H} NMR** (176 MHz, CDCl$_3$): δ = 159.3 (2x C=O minor), 159.2 (C^5C=O major), 157.5 (C^4C=O major), 142.5 ($C^{4/5}$ minor), 139.4 (d, $J_{C,F}$ = 0.9 Hz, C^5 major), 131.9 (C^4 major), 117.6 (q, $^1J_{C,F}$ = 272 Hz, CF_3 major), 54.9 (C^5COCH_3 major), 53.8 (2x CH_3 minor), 53.5 (C^4COCH_3 major), CF_3 (minor) not observed in ^{13}C{^{1}H}; **^{15}N NMR** (71 MHz, CDCl$_3$): δ = 248 (d, $J_{F,N}$ = 22 Hz, N^2 minor), 245 (d, $J_{F,N}$ = 21 Hz, N^1 major), 184 (N^2 major), 154 (N^1, N^3 minor); N^3 (major) not observed in ^{19}F^{15}N HMQC; **^{19}F NMR** (659 MHz, CDCl$_3$): δ = –57.1 (CF_3 major), -61.4 (CF_3 minor); **HRMS (EI)** calcd m/z for $C_7H_6N_3O_4F_3$: 253.0305 [M$^+$], found: 253.0302 [M$^+$];

Experimental Part

Elemental Analysis calcd (%) for $C_7H_6N_3O_4F_3$: C 33.21, H 2.39, N 16.60, O 25.28, F 22.52; found: C 33.18, H 2.63, N 16.24.

4,5-Diphenyl-1-(trifluoromethyl)-1H-1,2,3-triazole (89a) and 4,5-diphenyl-2-(trifluoromethyl)-2H-1,2,3-triazole (89b)

4,5-Diphenyl-1H-1,2,3-triazole (122 mg, 0.55 mmol, 1.1 equiv) was reacted with **1a** according to the general procedure and the resulting regioisomeric mixture was separated by flash chromatography (Alox B, act. I, pentane/Et$_2$O gradient, 1:0 to 20:1). Solvent was removed under reduced pressure and carefully dried under vacuum (0 °C, 10^{-3} mbar) to obtain **89a** as a white crystalline solid (30.4 mg, 21%) and **89b** as colorless oil containing traces of 4-Iodo-α-methylstyrene (15.9 mg, <11%). Single crystals for X-ray analysis of **89a** were obtained by sublimation (10^{-2} mbar, 60 °C).

(**89a**). m.p.: 68 °C; R_f (SiO$_2$, pentane/CH$_2$Cl$_2$ 1:1): 0.54; **^{1}H NMR** (500 MHz, CDCl$_3$): δ = 7.60-7.49 (m, 5H), 7.40-7.36 (m, 2H) 7.32-7.25 (m, 3H); **^{13}C{^{1}H} NMR** (126 MHz, CD$_2$Cl$_2$): δ = 146.0 (u q, $J_{C,F}$ = 2.0 Hz, C^5), 133.1 (C^4), 130.6 (C^5C_{p-Ph}), 130.1 (u q, $J_{C,F}$ = 0.5 Hz, 2x C^5C_{m-Ph}), 129.3 (2x C^4C_{m-Ph}), 129.2 ($C^4C_{ipso-Ph}$), 128.8 (C^4C_{p-Ph}), 128.7 (2x C^4C_{o-Ph}), 127.2 (2x C^5C_{o-Ph}), 125.5 (u q, $J_{C,F}$ = 0.5 Hz, $C^5C_{ipso-Ph}$), 118.1 (q, $^1J_{C,F}$ = 269 Hz, CF$_3$); **^{15}N NMR** (41 MHz, CD$_2$Cl$_2$): δ = 357 (N^2), 245 (d, $J_{F,N}$ = 14 Hz, N^1), N^3 not observed; **^{19}F NMR** (659 MHz, CDCl$_3$): δ = –55.4; **HRMS (EI)** calcd m/z for $C_{15}H_{10}F_3N_3$: 289.0822 [M$^+$], found: 289.0822 [M$^+$]; **Elemental Analysis** calcd (%) for $C_{15}H_{10}F_3N_3$: C 62.28, H 3.48, N 14.53, F 19.70; found: C 62.02, H 3.66, N 14.59; **CCDC**: 843429.

(**89b**). R_f (SiO$_2$, pentane/CH$_2$Cl$_2$ 5:1): 0.34; **^{1}H NMR** (500 MHz, CDCl$_3$): δ = 7.69-7.54 (m, 4H, CH_{o-Ph}), 7.53-7.37 (m, 6H, 4x CH_{m-Ph} / 2x CH_{p-Ph}); **^{13}C{^{1}H} NMR** (126 MHz, CDCl$_3$): δ = 148.6 (C^4, C^5) 135.0 (2x $C_{ipso-Ph}$), 129.7 (4x CH_{o-Ph}), 128.8 (4x CH_{m-Ph}), 128.6 (2x CH_{p-Ph}), 117.6 (d, $^1J_{C,F}$ = 267 Hz, CF$_3$); **^{15}N NMR** (41 MHz, CD$_2$Cl$_2$): δ = 325 (d, $J_{F,N}$ = 18.6 Hz, N^2), 241 (N^1, N^3); **^{19}F NMR** (659 MHz, CDCl$_3$): δ = –61.6; **HRMS (EI)** calcd m/z for $C_{15}H_{10}F_3N_3$: 289.0822 [M$^+$], found: 289.0823 [M$^+$].

5-Phenyl-2-(trifluoromethyl)-2H-tetrazole (90a) and 5-phenyl-1-(trifluoromethyl)-1H-tetrazole (90b)

5-Phenyl-1H-tetrazole (81 mg, 0.55 mmol, 1.1 equiv) was reacted with 2 according to the general procedure, without addition of HNTf$_2$ and performing the trifluoromethylation at room temperature instead of 35 °C. To facilitate purification, 3HF-NEt$_3$ (33 µL, 0.20 mmol, 0.4 equiv) was added after the reaction was complete. After stirring for 30 min at room temperature, saturated aqueous NaHCO$_3$ was added and the mixture was extracted (3x pentane). The combined organic phases were washed with brine, dried over MgSO$_4$ and the solvent removed under reduced pressure. The regioisomers were separated by flash chromatography (Alox N, act. I, pentane to pentane/Et$_2$O 1:1) to yield **90a** (19.7 mg, 18%) and **90b** (11.2 mg, 10%).

(**90a**). R_f (SiO$_2$, pentane): 0.3; **^{1}H NMR** (700 MHz, CDCl$_3$): δ = 8.26 (d, $J_{H,H}$ = 8.16 Hz, 2H, C$H_{o\text{-Ph}}$), 7.60-7.56 (m, 3H, C$H_{p\text{-Ph}}$ / C$H_{m\text{-Ph}}$); **^{13}C{^{1}H} NMR** (176 MHz, CDCl$_3$): δ = 166.7 (C^5), 131.7 ($C_{p\text{-Ph}}$), 129.2 (2x $C_{m\text{-Ph}}$), 127.5 (2x $C_{o\text{-Ph}}$), 125.4 ($C_{ipso\text{-Ph}}$), 117.0 (q, $^1J_{C,F}$ = 272 Hz, CF$_3$); **^{15}N NMR** (71 MHz, CDCl$_3$): δ = 279 ($J_{F,N}$ = 18 Hz, N^2), 185 (N); **^{19}F NMR** (659 MHz, CDCl$_3$): δ = – 60.7; **HRMS (EI)** calcd m/z for C$_8$H$_5$F$_3$N$_4$: 190.0713 [M$^+$], found: 190.0715 [M$^+$]; **Elemental Analysis** calcd (%) for C$_8$H$_5$F$_3$N$_4$: C 44.87, H 2.35, N 26.16, F 26.61; found: C 44.77, H 2.40, N 26.08, F 26.45.

(**90b**). R_f (SiO$_2$, pentane/Et$_2$O 10:1): 0.4; **^{1}H NMR** (700 MHz, CDCl$_3$): δ = 7.75 (d, $J_{H,H}$ = 7.32 Hz, 2H, C$H_{o\text{-Ph}}$), 7.69 (Ψt, $J_{H,H}$ = 7.86 Hz, 7.02 Hz, 1H, C$H_{p\text{-Ph}}$), 7.62 (Ψt, $J_{H,H}$ = 7.78 Hz, 7.71 Hz, 2H, C$H_{m\text{-Ph}}$); **^{13}C{^{1}H} NMR** (176 MHz, CDCl$_3$): δ = 154.1 (q, $J_{C,F}$ = 1.3 Hz, C^5), 132.5 ($C_{p\text{-Ph}}$), 129.2 (2x $C_{m\text{-Ph}}$), 129.2 (q, $J_{C,F}$ = 1.2 Hz, 2C, 2x $C_{o\text{-Ph}}$), 121.9 ($C_{ipso\text{-Ph}}$), 117.4 (q, $^1J_{C,F}$ = 273 Hz, CF$_3$); **^{15}N NMR** (71 MHz, CDCl$_3$): δ = 233 ($J_{F,N}$ = 17 Hz, N^1), 166 (N); **^{19}F NMR** (659 MHz, CDCl$_3$): δ = – 54.7; **HRMS (EI)** calcd m/z for C$_8$H$_5$F$_3$N$_4$: 190.0713 [M$^+$], found: 190.0715 [M$^+$]; **Elemental Analysis** calcd (%) for C$_8$H$_5$F$_3$N$_4$: C 44.87, H 2.35, N 26.16, F 26.61; found: C 44.93, H 2.36, N 25.98, F 26.37.

2-(Ethylthio)-1-(trifluoromethyl)-1*H*-benzo[*d*]imidazole (91)

The synthesis was carried out in a larger scale due to the tediousness of purification of the product. 2-Mercaptoethylbenzimidazole (294 mg, 1.65 mmol, 1.1 equiv) was reacted with **1a** (495 mg, 1.5 mmol) according to the general procedure and the resulting crude product was purified by flash chromatography (SiO$_2$, pentane/Et$_2$O 50:1), the product containing fractions were then concentrated under reduced pressure and further purified by chromatography (SiO$_2$, pentane/DCM 2:1). To obtain an analytical sample, the product fraction was heated to 50 °C and all volatile compounds were removed under reduced pressure (48 mg, 13%). R_f (pentane/Et$_2$O 50:1): 0.3; R_f (pentane/DCM 2/1): 0.24; **^{1}H NMR** (700 MHz, CD$_2$Cl$_2$): δ = 7.66 (d, $J_{H,H}$ = 7.7 Hz, 1H, C^4H), 7.54 (dm, $J_{H,H}$ = 7.7 Hz, 1H, C^7H), 7.36 (tm, $J_{H,H}$ = 7.7 Hz, 1H, C^6H), 7.32 (tm, $J_{H,H}$ = 7.7 Hz, 1H, C^5H), 3.43 (q, $J_{H,H}$ = 7.4 Hz, 2H, CH$_2$), 1.52 (t, $J_{H,H}$ = 7.4 Hz, 3H, CH$_3$). **^{13}C{^{1}H} NMR** (176 MHz,

CD$_2$Cl$_2$): δ = 151.1 (q, $J_{C,F}$ = 1.6 Hz, C^2), 143.3 (C^{7a}), 133.3 (C^{3a}), 124.1 (C^6), 123.6 (C^5), 119.2 (q, $^1J_{C,F}$ = 264 Hz, CF_3), 118.6 (C^4), 111.2 (q, $J_{C,F}$ = 4Hz, C^7), 26.6 (q, $J_{C,F}$ = 1.6 Hz, CH_2), 14.1 (CH_3); **^{15}N NMR** (71 MHz, CD$_2$Cl$_2$): δ = 153.2 (d, $J_{F,N}$ = 17 Hz, N^1), N^3 not observed; **^{19}F NMR** (658 MHz, CD$_2$Cl$_2$): δ = −54.5 (d, $J_{F,H}$ = 1.8 Hz, CF_3); **HRMS (EI)** calcd m/z for C$_{10}$H$_9$F$_3$N$_2$S: 246.0433 [M$^+$], found: 246.0432 [M$^+$]; **Elemental Analysis** calcd (%) for C$_{10}$H$_9$F$_3$N$_2$S: C 48.77, H 3.68, N 11.38, F 23.14, S 13.02; found: C 48.80, H 3.81, N 11.41, F 23.27, S 13.01.

2-Methyl-3-(trifluoromethyl)-1H-indole (92a, major) and 2-methyl-4-(trifluoromethyl)-1H-indole (92b, minor)

3-Methyl-1H-indole (154 mg, 0.55 mmol, 1.1 equiv) was reacted with **1a** according to the general procedure and crude product was purified by flash chromatography (SiO$_2$, pentane/Et$_2$O/DCM 5:1:1) and (SiO$_2$, pentane/Et$_2$O 5:1) to yield the title compounds **92a** and **92b** as a isomeric mixture containing minor impurities. **^{1}H NMR** (400 MHz, CD$_2$Cl$_2$): δ = 8.29 (br. s, NH), 7.68 (d, J = 7.5, 1H, C^{4}H major), 7.53 (d, J = 8.4 Hz, 1H, C^{7}H minor), 7.42-7.36 (m, 1H major / 1H minor, C^{6}H major / C^{5}H minor), 7.28-7.17 (m, 2H major / 1H minor, C^{5}H major/ C^{7}H major / C^{6}H minor), 6.43 (m, 1H, C^{3}H minor), 2.59 (q, $J_{H,F}$ = 1.7 Hz, 3H, CH_3 major), 2.53 (d, $J_{H,F}$ = 0.8 Hz, CH_3 minor); **^{13}C{^{1}H}C NMR** (101 MHz, CD$_2$Cl$_2$): δ = 137.9 (C^2 minor), 136.9 (C^2 major), 134.9 (C^{3a} major), 125.8 (q, $^1J_{C,F}$ = 267 Hz, CF_3 major), 125.8 (C^{7a} minor), 125.7 (q, $^1J_{C,F}$ = 272 Hz, CF_3 minor, derived from ^{19}F^{13}C HMQC), 125.5 (C^{7a} major), 122.8 (C^5H major), 121.3 (C^7H major), 120.2 (C^6H minor), 118.9 (C^4H major), 117.3 (q, $J_{C,F}$ = 4.9 Hz, C^5H minor), 114.2 (C^7H minor), 111.0 (C^6H major), 103.2 (q, $J_{C,F}$ = 3.6 Hz, C^3 major, derived from ^{19}F^{13}C HMQC), 99.6 (C^3H minor), 13.8 (CH_3 minor), 12.7 (CH_3 major), C^{3a} and C^4 minor not observed in ^{13}C spectra; **^{15}N NMR** (41 MHz, CD$_2$Cl$_2$): δ = 134 (N major), 131 (N minor); **^{19}F NMR** (377 MHz, CD$_2$Cl$_2$): δ = −54.9 (m, CF_3 major), −61.7 (CF_3 minor). **HRMS (EI)** calcd m/z for C$_{10}$H$_8$F$_3$N: 199.0604 [M$^+$], found: 199.0601 [M$^+$].

2-Methyl-4-(trifluoromethyl)-1H-indole (92b,minor) **CAS**: 1018971-85-1.

Ethyl 6,7,8-trifluoro-4-(trifluoromethoxy)quinoline-3-carboxylate (94)

Ethyl 6,7,8-trifluoroquinoline-3-carboxylate ((**93**), 154 mg, 0.55 mmol, 1.1 equiv) was reacted with **1a** according to the general procedure and resulting crude product purified by flash chromatography (SiO$_2$, pentane/Et$_2$O gradient 20/1 to 15:1) to yield the title compound as a colorless oil (40 mg, 24%). **m.p.**: 64 °C; R_f

(pentane/Et$_2$O 15:1): 0.2; **^{1}H NMR** (400 MHz, CD$_2$Cl$_2$): δ = 9.41 (s, 1H, C^{2}H), 7.86 (m, 1H, C^{5}H), 4.52 (q, $J_{H,H}$ = 7.2 Hz, 2H, CH$_2$), 1.47 (t, $J_{H,H}$ = 7.2 Hz, 3H, CH$_3$); **^{13}C{^{1}H} NMR** (101 MHz, CD$_2$Cl$_2$): δ = 163.0 (CO$_2$Et), 152.6 (C^{2}H), 151.9 (ddd, $J_{C,F}$ = 256 Hz, $J_{C,F}$ = 13 Hz, $J_{C,F}$ = 2.6 Hz, C^{6}F), 151.4 (m, C^{8a}), 146.4 (ddd, $J_{C,F}$ = 261 Hz, 9.8 Hz, 4.0 Hz, C^{8}F), 142.8 (ddd, $J_{C,F}$ = 260 Hz, $J_{C,F}$ = 19 Hz, $J_{C,F}$ = 13 Hz, C^{7}F), 140.2 (dm, $J_{C,F}$ = 10 Hz, C^{4a}), 120.6 (q, $J_{C,F}$ = 262 Hz, CF$_3$), 119.8 (C^3), 119.7 (d, $J_{C,F}$ = 9.5 Hz, C^{5}H), 104.3 (m, C^4), 63.0 (CH$_2$), 14.1 (CH$_3$); **^{19}F NMR** (376 MHz, CD$_2$Cl$_2$): δ = –56.7 (s, 3F, CF$_3$), –128.8 (ddd, J = 17.8 Hz, J = 10 Hz, J = 8 Hz, 1F, C^{6}F), –143.5 (ddd, J = 17.2 Hz, J = 8 Hz, J = 2.3 Hz, 1F, C^{8}F), –150.7 (ddd, J = 20.1 Hz, J =17.2 Hz, J = 7.5 Hz, 1F, C^{7}F); **HRMS (EI)** calcd m/z for C$_{13}$H$_7$F$_6$NO$_3$: 339.0325 [M$^+$], found: 339.0328 [M$^+$]; **Elemental Analysis** calcd (%) for C$_{13}$H$_7$F$_6$NO$_3$: C 46.03, H 20.8, N 4.13, O 14.15, F 33.61; found: C 46.21, H 2.17, N 4.10, F 33.69.

2,4,6-Trimethyl-3-(trifluoromethyl)phenol (95)

2,4,6-Trimethylphenol (77 mg, 0.55 mmol, 1.1 equiv) was reacted with **1a** according to the general procedure and the residue purified by flash chromatography (SiO$_2$, pentane/Et$_2$O 15:1) to yield the title compound as a colorless oil (28 mg, 27%) containing minor impurities. **^{1}H NMR** (400 MHz, CDCl$_3$): δ = 6.78 (s, 1H,. C^{5}H), 4.69 (s, 1H, OH), 2.40 (q, J = 4 Hz, 3H, C^{4}CH$_3$), 2.37 (q, J = 2.4 Hz, 3H, C^{2}CH$_3$), 2.26 (s, 3H, C^{6}CH$_3$); **^{13}C{^{1}H} NMR** (100.6 MHz, CDCl$_3$): δ = 151.0 (C^{1}OH), 132.4 (C^{5}H), 128.9 (C^4), 126.8 (C^3), 126.5 (C^6), 126.3 (q, $J_{C,F}$ = 276 Hz, CF$_3$), 21.5 (C^{4}CH$_3$), 16.4 (C^{6}CH$_3$), 12.9 (C^{2}CH$_3$). **^{19}F NMR** (282 MHz, CDCl$_3$): δ = –53.2; **HRMS (EI)** calcd m/z for C$_{10}$H$_{11}$F$_3$O: 204.0757 [M$^+$], found: 204.0758 [M$^+$].

5.4 Rate Study

5.4.1 Trifluoromethylation of *para*-Toluenesulfonic Acid Monohydrate

All reactions were monitored by ^{19}F NMR spectroscopy using a *Bruker* DPX 400 MHz NMR spectrometer operating at 376.5 MHz. Experimental temperatures (298 K) were maintained using a *Bruker* BVT3000 temperature control unit calibrated with a digital thermometer fit to a 5 mm NMR tube. Initially, the temperature in the spectrometer was equilibrated on a "standard" sample containing **1a** (600 µL, 0.1 M in CDCl$_3$/*t*BuOH 5:1) and internal standard (C$_6$H$_5$CF$_3$, 0.05 M) and the shim was optimized. A second tube charged with correct amount of the trifluoromethylating agent (300 µL, 0.2 M in CDCl$_3$/*t*BuOH 5:1 containing 0.05 M C$_6$H$_5$CF$_3$) and the appropriate amount of *para*-toluenesulfonic acid monohydrate (300 µL, 0.2 M in CDCl$_3$/*t*BuOH 5:1 containing 0.05 M C$_6$H$_5$CF$_3$) was added. The tube was shaken vigorously for 10–15 seconds after which it was exchanged with the standard in the spectrometer and the acquisition program was started.

Acquisition program: A pseudo 2D NMR experiment designed for kinetic measurements was utilized to monitor reaction progress for all rate experiments. The time interval between individual acquisitions (d1), number of acquisitions averaged (one data point was obtained as the average of 2 individual acquisitions) and total number of data points could be varied to effect the frequency of acquisition, signal to noise ratio and experiment duration, respectively.

Data Processing: For each data point, integrals corresponding to the ^{19}F NMR resonances of the trifluoromethyl group of the trifluoromethylating reagent, the trifluoromethyl group of the newly formed trifluoromethyl *para*-toluenesulfonic acid ester and C$_6$H$_5$CF$_3$ internal standard were extracted using Bruker's XWinNMR 3.5 software package. The resulting data were exported as tables of integral values for each signal over all data points measured. The ^{19}F NMR integration data were then imported into SigmaPlot10 and the data point numbers transformed into time values by multiplying by the correct acquisition duration and number of acquisitions averaged per data point. The time vs. ^{19}F NMR integration data thus generated were fit to [c] = $v_0 t + b$ for a linear increase to a maximum of 10% conversion in order to extract initial rate values (v_0).

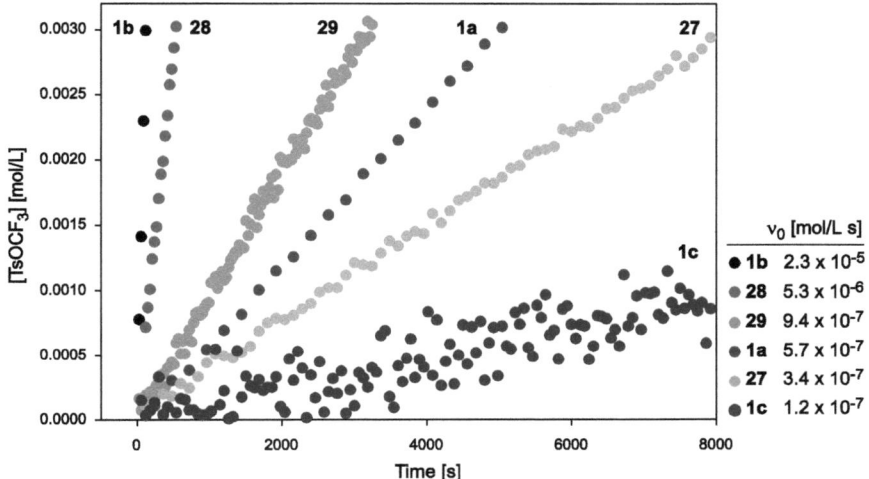

Scheme 26. Profile for the trifluoromethylation of *para*-toluenesulfonic acid monohydrate (0.1 M) with different trifluoromethylating agent monitored by ^{19}F NMR and corresponding initial rates constants.

5.4.2 Formation of (E)-N-(1-(1H-Benzo[d][1,2,3]triazol-1-yl)ethyliden)trifluoromethanamine (39)

All reactions were monitored by ^{19}F NMR spectroscopy using a *Bruker* DPX 400 MHz NMR spectrometer operating at 376.5 MHz. Experimental temperature (333 K) was maintained using a *Bruker* BVT 3000 temperature control unit calibrated with a digital thermometer fit to a 5 mm NMR tube. Initially, the temperature in the spectrometer was equilibrated on a "standard" sample containing benzotriazole (0.076 mmol, 0.15 M in CD$_3$CN) and the internal standard (C$_6$H$_5$CF$_3$, 0.05 mmol) and the shim was optimized. A second tube charged with benzotriazole (0.076 mmol, 0.15 M in CD$_3$CN), HNTf$_2$ (0.0035 mmol, 0.1 M in CH$_2$Cl$_2$), the trifluoromethylating agent **1a** (0.052 mmol, 0.1 M in CD$_3$CN) and the internal standard (C$_6$H$_5$, 0.05 mmol) was added. The tube was shaken vigorously for 10–15 seconds after which it was exchanged with the standard in the spectrometer and the acquisitionprogram was started.

Acquisition program: A pseudo 2D NMR experiment designed for kinetic measurements was utilized to monitor the reaction progress for the rate experiment. The time interval between individual acquisitions (d1), number of acquisitions averaged (one data point was obtained as the average of 4 individual acquisitions) and total number of data points could be varied to effect the frequency of acquisition, signal to noise ratio and experiment duration, respectively.

Experimental Part

Data Processing: For each data point, integrals corresponding to the ^{19}F NMR resonances of the trifluoromethyl group of the trifluoromethylating reagent, the trifluoromethyl group of the newly formed N-substituted N-trifluoroimine, the trifluoromethyl group of the two side products and $C_6H_5CF_3$ as internal standard were extracted using *Bruker*'s XWin NMR 3.5 software package. The resulting data were exported as tables of integral values for each signal over all data points measured. The ^{19}F NMR integration data were then imported into SigmaPlot10 and the data point numbers transformed into time values by multiplying by the correct acquisition duration and number of acquisitions averaged per data point.

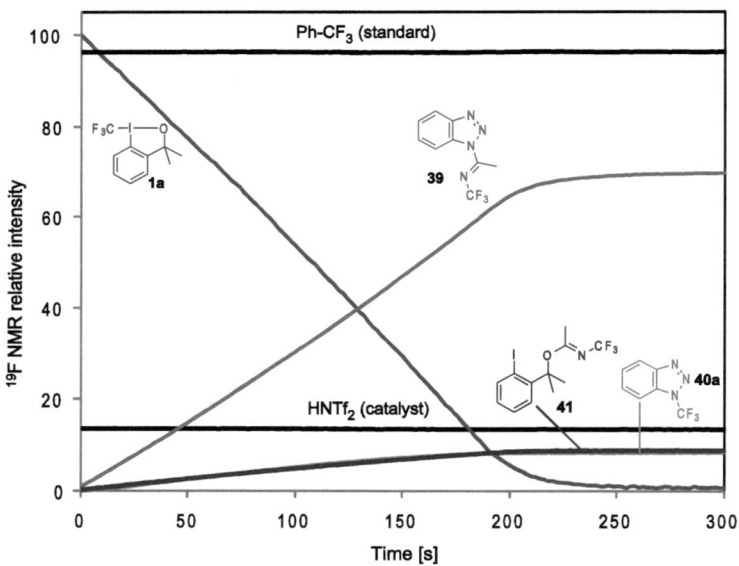

Scheme 27. Reaction profile for the formation of (*E*)-*N*-(1-(1H-Benzo[*d*][1,2,3]triazol-1-yl)-ethyliden)trifluoromethanamine (**39**).

5.4.3 *N*-Trifluoromethylation of 1-(Trimethylsilyl)-1*H*-benzo[*d*]triazole (65)

All reactions were monitored by ^{19}F NMR spectroscopy using a *Bruker* DRX 400 MHz NMR spectrometer operating at 376.5 MHz. The experimental temperature (308 K) was maintained using a *Bruker* BVT 3000 temperature control unit calibrated with a digital thermometer, fit to a 5 mm NMR tube. Initially, the temperature in the spectrometer was equilibrated on a "standard" sample containing **65** (0.50 mmol, 1.65 M in CD_2Cl_2), reagent **1a** (0.45 mmol, 1.5 M in CD_2Cl_2) and an internal standard ($C_6H_5CF_3$, 0.29 mmol) and the shim was optimized. To a second tube charged with **65** (0.50 mmol, 1.65 M in CD$_2$Cl$_2$), the trifluoromethylating agent **1a** (0.45 mmol, 1.5 M in CD$_2$Cl$_2$) and the internal standard (C$_6$H$_5$CF$_3$, 0.29 mmol) was added HNTf$_2$ (0.045 mmol, 0.5 M in CH$_2$Cl$_2$) or BF$_3$-Et$_2$O (0.023 mmol). The tube was shaken vigorously for 10–15 seconds whereupon it was exchanged with the standard in the spectrometer and the acquisition program was started.

Acquisition program: A pseudo 2D NMR experiment designed for kinetic measurements was utilized to monitor the reaction progress for the rate experiment. The time interval between individual acquisitions (d1) was set to 117 seconds resulting in one complete acquisition (1 scan) every two minutes (120s).

Data Processing: For each data point, integrals corresponding to all ^{19}F NMR resonances of interest and PhCF$_3$ internal standard were extracted using *Bruker's* XWin NMR 3.5 software package. The resulting data were exported as tables of integral values for each signal over all data points measured. The ^{19}F NMR integration data were then imported into Microsoft Excel 2007 and then normalized to the maximum integral value observed for plotting.

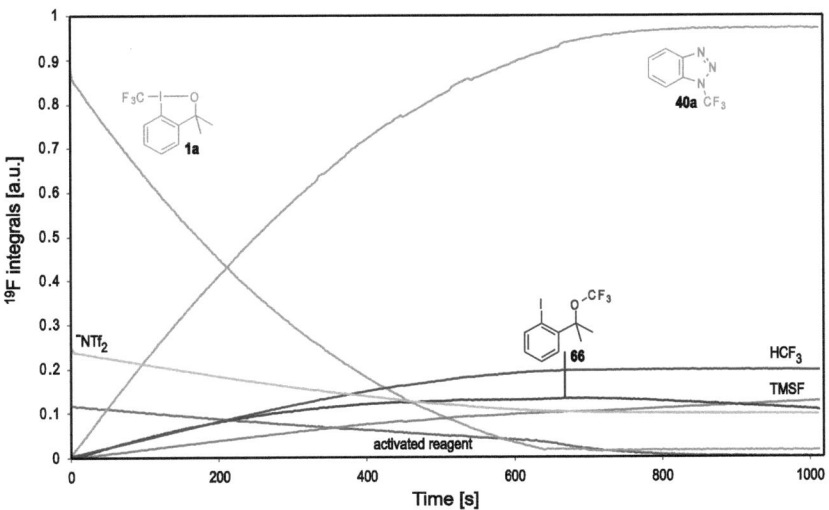

Scheme 28. Reaction profile of *N*-trifluoromethylation of 1-(trimethylsilyl)benzo[*d*]triazole using 12 mol-% HNTf$_2$ acid catalyst.

Experimental Part

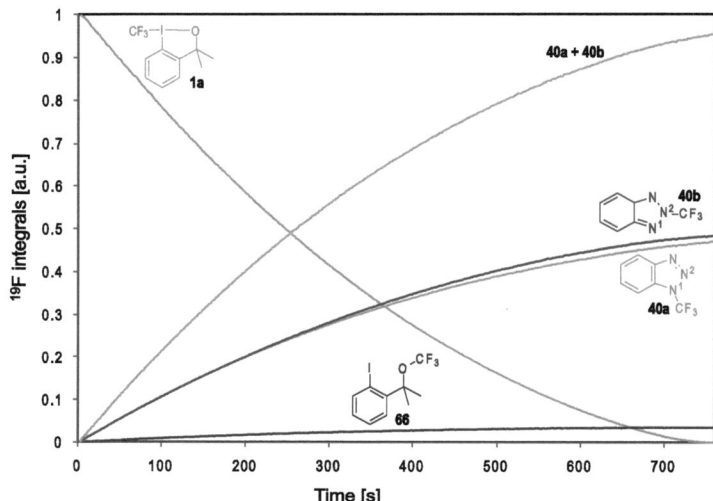

Scheme 29. Reaction profile of N-trifluoromethylation of 1-(trimethylsilyl)benzo[d]triazole using 5 mol-% BF_3-Et_2O acid catalyst.

6 Literature

[1] X.-H. Xu, G.-M. Yao, Y.-M. Li, J.-H. Lu, C.-J. Lin, X. Wang, C.-H. Kong, *J. Nat. Prod.* **2003**, *66*, 285-288.

[2] *CRC Handbook of Chemistry and Physics, 92nd Edition*, CRC Press, **2011**.

[3] N. Shibata, A. Matsnev, D. Cahard, *Beilstein J. Org. Chem.* **2010**, *6*, 65.

[4] A. M. Thayer, *Chem. Eng. News* **2006**, *84*, 15-24.

[5] M. A. McClinton, D. A. McClinton, *Tetrahedron* **1992**, *48*, 6555-6666.

[6] H.-J. Böhm, D. Banner, S. Bendels, M. Kansy, B. Kuhn, K. Müller, U. Obst-Sander, M. Stahl, *ChemBioChem* **2004**, *5*, 637-643.

[7] P. Eisenberger, Diss. ETH No. 17371, ETH (Zürich), **2007**.

[8] W. R. Hasek, W. C. Smith, V. A. Engelhardt, *J. Am. Chem. Soc.* **1960**, *82*, 543-551.

[9] a) G. S. Lal, G. P. Pez, R. J. Pesaresi, F. M. Prozonic, *Chem. Commun.* **1999**, 215-216; b) G. S. Lal, G. P. Pez, R. J. Pesaresi, F. M. Prozonic, H. S. Cheng, *J. Org. Chem.* **1999**, *64*, 7048-7054.

[10] W. J. Middleton (E. I. Du Pont de Nemours & Company), US-3914265, **1975**.

[11] T. Umemoto, R. P. Singh, Y. Xu, N. Saito, *J. Am. Chem. Soc.* **2010**, *132*, 18199-18205.

[12] a) M. Kuroboshi, T. Hiyama, *Tetrahedron Lett.* **1992**, *33*, 4177-4178; b) T. Hiyama, K. Kanie, K. Mizuno, M. Kuroboshi, *Bull. Chem. Soc. Jpn.* **1998**, *71*, 1973-1991; c) M. Kuroboshi, K. Kanie, T. Hiyama, *Adv. Synth. Catal.* **2001**, *343*, 235-250.

[13] S. Rozen, *Acc. Chem. Res.* **2005**, *38*, 803-812.

[14] J.-A. Ma, D. Cahard, *J. Fluorine Chem.* **2007**, *128*, 975-996.

[15] G. G. Dubinina, H. Furutachi, D. A. Vicic, *J. Am. Chem. Soc.* **2008**, *130*, 8600-8601.

[16] a) M. Oishi, H. Kondo, H. Amii, *Chem. Commun.* **2009**; b) E. J. Cho, T. D. Senecal, T. Kinzel, Y. Zhang, D. A. Watson, S. L. Buchwald, *Science* **2010**, *328*, 1679-1681.

[17] T. D. Senecal, A. T. Parsons, S. L. Buchwald, *J. Org. Chem.* **2011**, *76*, 1174-1176.

[18] L. Chu, F.-L. Qing, *J. Am. Chem. Soc.* **2011**, *134*, 1298-1304.

[19] X. Mu, S. J. Chen, X. L. Zhen, G. S. Liu, *Chem. - Eur. J.* **2011**, *17*, 6039-6042.

Literature

[20] E. J. Cho, S. L. Buchwald, *Org. Lett.* **2011**, *13*, 6552-6555.

[21] X. Mu, T. Wu, H.-y. Wang, Y.-l. Guo, G. Liu, *J. Am. Chem. Soc.* **2011**, *134*, 878-881.

[22] L. Chu, F.-L. Qing, *J. Am. Chem. Soc.* **2010**, *132*, 7262-7263.

[23] I. Ruppert, K. Schlich, W. Volbach, *Tetrahedron Lett.* **1984**, *25*, 2195-2198.

[24] G. K. S. Prakash, R. Krishnamurti, G. A. Olah, *J. Am. Chem. Soc.* **1989**, *111*, 393-395.

[25] a) G. K. S. Prakash, A. K. Yudin, *Chem. Rev.* **1997**, *97*, 757-786; b) G. K. S. Prakash, M. Mandal, *J. Fluorine Chem.* **2001**, *112*, 123-131.

[26] G. Pawelke, *J. Fluorine Chem.* **1989**, *42*, 429-433.

[27] G. K. S. Prakash, J. B. Hu, G. A. Olah, *J. Org. Chem.* **2003**, *68*, 4457-4463.

[28] S. Ait-Mohand, N. Takechi, M. Medebielle, W. R. Dolbier, *Org. Lett.* **2001**, *3*, 4271-4273.

[29] G. K. S. Prakash, J. B. Hu, G. A. Olah, *Org. Lett.* **2003**, *5*, 3253-3256.

[30] a) G. Blond, T. Billard, B. R. Langlois, *Tetrahedron Lett.* **2001**, *42*, 2473-2475; b) T. Billard, B. R. Langlois, *J. Org. Chem.* **2002**, *67*, 997-1000; c) T. Billard, B. R. Langlois, G. Blond, *Eur. J. Org. Chem.* **2001**, 1467-1471.

[31] a) S. Roussel, T. Billard, B. R. Langlois, L. Saint-Jalmes, *Synlett* **2004**, 2119-2122; b) S. Roussel, T. Billard, B. R. Langlois, L. Saint-James, *Chem. - Eur. J.* **2005**, *11*, 939-944.

[32] W. R. Dolbier, *Chem. Rev.* **1996**, *96*, 1557-1584.

[33] Y. Zheng, J. A. Ma, *Adv. Synth. Catal.* **2010**, *352*, 2745-2750.

[34] Y. Itoh, K. Mikami, *Tetrahedron* **2006**, *62*, 7199-7203.

[35] D. A. Nagib, M. E. Scott, D. W. C. MacMillan, *J. Am. Chem. Soc.* **2009**, *131*, 10875-10877.

[36] Y. Macé, E. Magnier, *Eur. J. Org. Chem.* **2012**, *2012*, 2479.

[37] R. Koller, Diss. ETH No. 19219, ETH (Zürich), **2010**.

[38] L. M. Yagupolskii, N. V. Kondratenko, G. N. Timofeeva, *J. Org. Chem. USSR* **1984**, *20*, 103-106.

[39] L. M. Yagupolskii, A. V. Matsnev, R. K. Orlova, B. G. Deryabkin, Y. L. Yagupolskii, *J. Fluorine Chem.* **2008**, *129*, 131-136.

[40] T. Umemoto, S. Ishihara, *J. Am. Chem. Soc.* **1993**, *115*, 2156-2164.

[41] J.-J. Yang, R. L. Kirchmeier, J. n. M. Shreeve, *J. Org. Chem.* **1998**, *63*, 2656-2660.

[42] a) E. Magnier, J. C. Blazejewski, M. Tordeux, C. Wakselman, *Angew. Chem. Int. Ed.* **2006**, *45*, 1279-1282; b) Y. Mace, B. Raymondeau, C. Pradet, J. C. Blazejewski, E. Magnier, *Eur. J. Org. Chem.* **2009**, 1390-1397.

[43] T. Umemoto, S. Ishihara, *Tetrahedron Lett.* **1990**, *31*, 3579-3582.

[44] T. Umemoto, *Chem. Rev.* **1996**, *96*, 1757-1778.

[45] X. S. Wang, L. Truesdale, J. Q. Yu, *J. Am. Chem. Soc.* **2010**, *132*, 3648.

[46] J. Xu, D. F. Luo, B. Xiao, Z. J. Liu, T. J. Gong, Y. Fu, L. Liu, *Chem. Commun.* **2011**, *47*, 4300-4302.

[47] J. Xu, Y. Fu, D.-F. Luo, Y.-Y. Jiang, B. Xiao, Z.-J. Liu, T.-J. Gong, L. Liu, *J. Am. Chem. Soc.* **2011**, *133*, 15300-15303.

[48] T. Umemoto, K. Adachi, *J. Org. Chem.* **1994**, *59*, 5692-5699.

[49] S. Noritake, N. Shibata, Y. Nomura, Y. Y. Huang, A. Matsnev, S. Nakamura, T. Toru, D. Cahard, *Org. Biomol. Chem.* **2009**, *7*, 3599-3604.

[50] T. Umemoto, K. Adachi, S. Ishihara, *J. Org. Chem.* **2007**, *72*, 6905-6917.

[51] A. Matsnev, S. Noritake, Y. Nomura, E. Tokunaga, S. Nakamura, N. Shibata, *Angew. Chem. Int. Ed.* **2010**, *49*, 572-576.

[52] K. Adachi, S. Ishihara (DAIKIN IND LTD), JP-2003000388769, **2005**.

[53] C. R. Johnson, E. R. Janiga, M. Haake, *J. Am. Chem. Soc.* **1968**, *90*, 3890-3891.

[54] a) I. Kieltsch, P. Eisenberger, A. Togni, *Angew. Chem. Int. Ed.* **2007**, *46*, 754-757; b) P. Eisenberger, S. Gischig, A. Togni, *Chem. - Eur. J.* **2006**, *12*, 2579-2586; c) M. S. Wiehn, E. V. Vinogradova, A. Togni, *J. Fluorine Chem.* **2010**, *131*, 951-957; d) V. Matoušek, A. Togni, V. Bizet, D. Cahard, *Org. Lett.* **2011**, *13*, 5762-5765; e) E. Mejía, A. Togni, *ACS Catal.* **2012**, 521-527.

[55] a) S. Capone, I. Kieltsch, O. Flögel, G. Lelais, A. Togni, D. Seebach, *Helv. Chim. Acta* **2008**, *91*, 2035-2056; b) N. Santschi, A. Togni, *J. Org. Chem.* **2011**, *76*, 4189-4193.

[56] a) P. Eisenberger, I. Kieltsch, N. Armanino, A. Togni, *Chem. Commun.* **2008**, 1575-1577; b) A. Sondenecker, J. Cvengros, R. Aardoom, A. Togni, *Eur. J. Org. Chem.* **2011**, 78-87; c) N. Armanino, R. Koller, A. Togni, *Organometallics* **2010**, *29*, 1771-1777.

[57] a) K. Stanek, R. Koller, A. Togni, *J. Org. Chem.* **2008**, *73*, 7678-7685; b) R. Koller, Q. Huchet, P. Battaglia, J. M. Welch, A. Togni, *Chem. Commun.* **2009**, 5993-5995; c) R. Koller, K. Stanek, D. Stolz, R. Aardoom, K. Niedermann, A. Togni, *Angew. Chem. Int. Ed.* **2009**, *48*, 4332-4336; d) N. Santschi, P.

Geissbühler, A. Togni, *J. Fluorine Chem.* **2012**, *135*, 83-86; e) S. Fantasia, J. M. Welch, A. Togni, *J. Org. Chem.* **2010**, *75*, 1779-1782.

[58] a) A. T. Parsons, S. L. Buchwald, *Angew. Chem. Int. Ed.* **2011**, *50*, 9120-9123; b) X. Wang, Y. Ye, S. Zhang, J. Feng, Y. Xu, Y. Zhang, J. Wang, *J. Am. Chem. Soc.* **2011**, *133*, 16410-16413.

[59] Z. Weng, H. Li, W. He, L.-F. Yao, J. Tan, J. Chen, Y. Yuan, K.-W. Huang, *Tetrahedron* **2012**, *68*, 2527-2531.

[60] T. F. Liu, Q. L. Shen, *Org. Lett.* **2011**, *13*, 2342-2345.

[61] R. Shimizu, H. Egami, T. Nagi, J. Chae, Y. Hamashima, M. Sodeoka, *Tetrahedron Lett.* **2010**, *51*, 5947-5949.

[62] A. T. Parsons, T. D. Senecal, S. L. Buchwald, *Angew. Chem. Int. Ed.* **2012**, *51*, 2947-2950.

[63] T. Liu, X. Shao, Y. Wu, Q. Shen, *Angew. Chem. Int. Ed.* **2012**, *51*, 540-543.

[64] A. E. Allen, D. W. C. MacMillan, *J. Am. Chem. Soc.* **2010**, *132*, 4986.

[65] J. L. Gay-Lussac, *Ann. Chim. Phys., Sér. 1* **1814**, *91*, 5.

[66] J. I. Musher, *Angew. Chem. Int. Ed. Engl.* **1969**, *8*, 54-68.

[67] C. Willgerodt, *J. Prakt. Chem.* **1886**, *33*, 154-160.

[68] D. B. Dess, J. C. Martin, *J. Org. Chem.* **1983**, *48*, 4155-4156.

[69] T. Wirth, *Angew. Chem. Int. Ed.* **2005**, *44*, 3656-3665.

[70] L. M. Yagupolskii, I. I. Maletina, N. V. Kondratenko, V. V. Orda, *Synthesis* **1978**, 835.

[71] V. V. Zhdankin, P. J. Stang, *Chem. Rev.* **2002**, *102*, 2523-2584.

[72] H. Yamataka, K. Yamaguchi, T. Takatsuka, T. Hanafusa, *Bull. Chem. Soc. Jpn.* **1992**, *65*, 1157-1158.

[73] G. A. Rabah, G. F. Koser, *Tetrahedron Lett.* **1996**, *37*, 6453-6456.

[74] T. Irie, H. Tanida, *J. Org. Chem.* **1978**, *43*, 3274-3277.

[75] A. Baranowski, D. Plachta, L. Skulski, M. Klimaszewska, *J. Chem. Res., Synop.* **2000**, 435-437.

[76] A. Podgorsek, M. Jurisch, S. Stavber, M. Zupan, J. Iskra, J. A. Gladysz, *J. Org. Chem.* **2009**, *74*, 3133-3140.

[77] A. A. Kolomeitsev, M. Vorobyev, H. Gillandt, *Tetrahedron Lett.* **2008**, *49*, 449-454.

[78] N. Ignatyev, W. Hierse, M. Seidel, A. Bathe, J. Schroeter, K. Koppe, T. Meier, P. Barthen, W. Frank (MERCK), US-2011/0082312 A1, **2011**.

[79] a) M. E. Redwood, C. J. Willis, *Can. J. Chem.* **1965**, *43*, 1893-1898; b) W. B. Farnham, B. E. Smart, W. J. Middleton, J. C. Calabrese, D. A. Dixon, *J. Am. Chem. Soc.* **1985**, *107*, 4565-4567.

[80] R. Minkwitz, R. Brochler, *Z. Naturforsch. B* **1997**, *52*, 401-404.

[81] a) C. Jortay, R. Flammang, A. Maquestiau, *Bull. Soc. Chim. Belg.* **1985**, *94*, 727-734; b) C. A. Deakyne, M. Meot-Ner, *J. Am. Chem. Soc.* **1999**, *121*, 1546-1557.

[82] R. Minkwitz, M. Berkei, R. Ludwig, *Eur. J. Inorg. Chem.* **2000**, *2000*, 2387-2392.

[83] R. Minkwitz, R. Bröchler, M. Schütze, *Z. Anorg. Allg. Chem.* **1995**, *621*, 1727-1730.

[84] ConQuest 1.13 Search of CSD version 5.32 updates November 2011, on 28.02.2012, for (Cl)IXAr X = heteroatom excl. Cl

[85] M. Takahashi, H. Nanba, T. Kitazawa, M. Takeda, Y. Ito, *J. Coord. Chem.* **1996**, *37*, 371-378.

[86] D. G. Naae, J. Z. Gougoutas, *J. Org. Chem.* **1975**, *40*, 2129-2131.

[87] T. M. Balthazor, D. E. Godar, B. R. Stults, *J. Org. Chem.* **1979**, *44*, 1447-1449.

[88] Thomson Reuters Integrity http://integrity.thomson-pharma.com, 07.03.2011

[89] ConQuest 1.13 Search of CSD version 5.32 updates November 2011, on 20.02.2012, for NCF_3

[90] G. Klöter, W. Lutz, K. Seppelt, W. Sundermeyer, *Angew. Chem. Int. Ed. Engl.* **1977**, *16*, 707-708.

[91] Y. Asahina, I. Araya, K. Iwase, F. Iinuma, M. Hosaka, T. Ishizaki, *J. Med. Chem.* **2005**, *48*, 3443-3446.

[92] W. Dmowski, M. Kaminski, *J. Fluorine Chem.* **1983**, *23*, 207-218.

[93] a) R. J. Harder, W. C. Smith, *J. Am. Chem. Soc.* **1961**, *83*, 3422-3424; b) L. N. Markovskij, V. E. Pashinnik, A. V. Kirsanov, *Synthesis* **1973**, 787.

[94] E. Klauke, *Angew. Chem. Int. Ed. Engl.* **1966**, *5*, 848.

[95] L. M. Yagupol'skii, N. V. Kondratenko, G. N. Timofeeva, M. I. Dronkina, Y. L. Yagupol'skii, *J. Gen. Chem. USSR Engl. Transl.* **1981**, *16*, 2139.

[96] G. Pawelke, *J. Fluorine Chem.* **1991**, *52*, 229-234.

[97] T. Abe, E. Hayashi, H. Baba, H. Fukaya, *J. Fluorine Chem.* **1990**, *48*, 257-279.

[98] Y. Hagooly, J. Gatenyo, A. Hagooly, S. Rozen, *J. Org. Chem.* **2009**, *74,* 8578-8582.

[99] I. Kieltsch, Diss. ETH No. 17990, ETH (Zürich), **2008**.

[100] I. A. Koppel, R. W. Taft, F. Anvia, S.-Z. Zhu, L.-Q. Hu, K.-S. Sung, D. D. DesMarteau, L. M. Yagupolskii, Y. L. Yagupolskii, *J. Am. Chem. Soc.* **1994**, *116,* 3047-3057.

[101] I. L. Knunyants, B. L. Dyatkin, *Russ. Chem. Bull.* **1964**, *13,* 863-865.

[102] J. Foropoulos, D. D. DesMarteau, *Inorg. Chem.* **1984**, *23,* 3720-3723.

[103] a) J. L. Kurz, J. M. Farrar, *J. Am. Chem. Soc.* **1969**, *91,* 6057-6062; b) D. J. Bowden, S. L. Clegg, P. Brimblecombe, *Chemosphere* **1996**, *32,* 405-420.

[104] E. M. Burgess, H. R. Penton, E. A. Taylor, *J. Org. Chem.* **1973**, *38,* 26-31.

[105] A. R. Katritzky, K. Yannakopoulou, *Heterocycles* **1989**, *28,* 1121-1134.

[106] J. Baumanns, L. Deneken, D. Naumann, M. Schmeisser, *J. Fluorine Chem.* **1974**, *3,* 323-327.

[107] a) H. Pinto des Magalhaes, MSc thesis, ETH (Zürich), **2011**; b) O. Sala, MSc thesis, ETH (Zürich), **2012**.

[108] a) H. Bähme, G. Braun, *Arch. Pharm.* **1984**, *317,* 411-417; b) R. Tiollais, *Bull. Soc. Chim. Fr.* **1947**, *14,* 716-724.

[109] L. M. Yagupolskii, D. V. Fedyuk, K. I. Petko, V. I. Troitskaya, V. I. Rudyk, V. V. Rudyuk, *J. Fluorine Chem.* **2000**, *106,* 181-187.

[110] D. H. Obrien, C. P. Hrung, *J. Organomet. Chem.* **1971**, *27,* 185-193.

[111] M. N. S. Rad, A. Khalafi-Nezhad, M. Divar, S. Behrouz, *Phosphorus, Sulfur Silicon Relat. Elem.* **2010**, *185,* 1943-1954.

[112] M. A. Zolfigol, *Tetrahedron* **2001**, *57,* 9509-9511.

[113] U. Domańska, A. Pobudkowska, M. Rogalski, *J. Chem. Eng. Data* **2004**, *49,* 1082.

[114] N. Sunduru, L. Gupta, K. Chauhan, N. N. Mishra, P. K. Shukla, P. M. S. Chauhan, *Eur. J. Med. Chem.* **2011**, *46,* 1232-1244.

[115] Star Wars II: Attack of the Clones; Dir. George Lukas, 20th Century Fox, **2002**.

[116] a) K. Niedermann, J. M. Welch, R. Koller, J. Cvengros, N. Santschi, P. Battaglia, A. Togni, *Tetrahedron* **2010**, *66,* 5753-5761; b) K. Niedermann, N. Fruh, E. Vinogradova, M. S. Wiehn, A. Moreno, A. Togni, *Angew. Chem. Int. Ed.* **2011**, *50,* 1059-1063; c) K. Niedermann, N. Früh, R. Senn, B. Czarniecki, R. Verel, A. Togni, *Angew. Chem. Int. Ed.* **2012**, *51,* 6511-6515.

[117] 6.02 ed., Bruker AXS, Madison, WI, **2001**.

[118] G. Sheldrick, *Acta Crystallogr., Sect. A* **1990**, *46,* 467-473.

[119] G. M. Sheldrick, Universität Göttingen, Göttingen, Germany, **1999**.

[120] R. Blessing, *Acta Crystallogr., Sect. A* **1995**, *51,* 33-38.

[121] A. Linden, *Acta Crystallogr., Sect. C* **2009**, *65,* e1-e2.

[122] M. C. Harsanyi, P. A. Lay, R. K. Norris, P. K. Witting, *Aust. J. Chem.* **1996**, *49,* 581-597.

[123] P. Eisenberger, I. Kieltsch, R. Koller, K. Stanek, A. Togni, *Org. Synth.* **2011**, *88,* 168.

[124] R. J. Koshar, L. L. Barber, (Minnesota Mining and Manufacturing Company), US-4053519, **1977**.

[125] R. M. Claramunt, C. Lopez, M. d. l. A. Garcia, M. Pierrot, M. Giorgi, J. Elguero, *J. Chem. Soc., Perkin Trans. 2* **2000**, 2049-2053.

[126] A. L. Rheingold, C. B. White, S. Trofimenko, *Inorg. Chem.* **1993**, *32,* 3471-3477.

[127] J. L. Huppatz, *Aust. J. Chem.* **1983**, *36,* 135-147.

[128] A. Echevarría, J. Elguero, *Synth. Commun.* **1993**, *23,* 925-930.

[129] A. V. Lesiv, S. L. Ioffe, Y. A. Strelenko, V. A. Tartakovsky, *Helv. Chim. Acta* **2002**, *85,* 3489-3507.

[130] U. Burckhardt, Diss. ETH No. 12167, ETH (Zürich), **1997**.

[131] E. F. Perozzi, R. S. Michalak, G. D. Figuly, W. H. Stevenson, D. B. Dess, M. R. Ross, J. C. Martin, *J. Org. Chem.* **1981**, *46,* 1049-1053.

[132] L. Birkofer, P. Wegner, *Chem. Ber.* **1966**, *99,* 2512-2517.

[133] K. Morimoto, K. Makino, S. Yamamoto, G. Sakata, *J. Heterocycl. Chem.* **1990**, *27,* 807-810.

Literature

7 Appendix

7.1 Abbreviations

AAC	Azide-alkyne cycloaddition
act.	Activity
Ad	Adamantyl
approx.	Approximatelly
Ar	Aryl
Bn	Benzyl
b.p.	Boiling point
Bu	Butyl
calcd	calculated
CAS	Chemical abstrats service
Cy	Cyclohexyl
DAST	Diethylaminosulfur trifluoride
DBH	1,3-Dibromo-5,5-dimethylhydantoin
DCE	1,2-Dichloroethan
DCM	Dichloromethane
dec.	Decomposition
det.	Determined
DMP	Dess-Martin periodane
dppe	1,2-Bis(diphenylphosphino)ethane
DSC	Differential scanning calorimetry
EA	Elemental analysis
Et	Ethyl
equiv	Equivalent
GC	Gas chromatography
HMBC	Heteronuclear multiple-bond correlation
HMDS	1,1,1,3,3,3-Hexamethyldisilazane
HMQC	Heteronuclear multiple-quantum correlation
*m*CPBA	*meta*-Chlorperbenzoesäure
Me	Methyl
Mes	Mesityl, 2,4,6-trimethylphenyl
m.p.	Melting point
MS	Mass spectrometry
na	not available
NBS	*N*-Bromosuccinimide

Appendix

NHC	N-Heterocyclic carbene, Arduengo carbene
NIS	*N*-Iodosuccinimide
NMR	Nuclear magnetic resonance
OAc	Acetate
PGSE	Pulsed-field gradient spin-echo
Ph	Phenyl
py	Pyridine
quant	quantitative
R_f	Retention factor
r.t.	Room temperature
R.E.	Reductive elimination
SSA	silica sulfuric acid
TBA	Tetrabutylammonium
TBAT	Tetrabutylammonium difluorotriphenylsilicate (IV)
TDAE	Tetrakis(dimethylamino)ethylene
Tf	Trifluoromethanesulfonyl
TFA	Trifluoroacetic acid
THF	Tetrahydrofuran
TMEDA	*N,N,N',N'*-Tetramethylethylenediamine
TMS	Trimethylsilyl
VSEPR	Valence shell electron pair repulsion

7.2 Crystallographic Data

1-Chloro-1,3-dihydro-3,3-dimethyl-1,2-benziodoxole (2a)

CCDC	771236
Empirical formula	$C_9H_{10}ClIO$
Formula weight	296.52
Temperature	100(2) K
Wavelength	0.71073 Å
Crystal system, space group	Triclinic, $P\bar{1}$
Unit cell dimensions	a = 8.0494(11) Å α = 91.061(2)°
	b = 8.0836(11) Å β = 106.358(2)°
	c = 8.8897(12) Å γ = 114.673(2)°
Volume	498.17(12) Å^3
Z, Calculated density	2, 1.977 Mg/m^3
Absorption coefficient	3.433 mm^{-1}
F(000)	284
Crystal size	0.709 x 0.232 x 0.174 mm
Data collection	Bruker SMART APEX platform with CCD Detector Graphite monochromator
Detector distance	50 mm
Method; exposure time/frame	omega-scans; t = 0.5 sec
Solution by	direct methods
Refinement method	full matrix least-squares on F^2, SHELXTL
Theta range for data collection	2.42 to 28.34°
Limiting indices	-10<=h<=10, -10<=k<=10, -11<=l<=11
Reflections collected / unique	5073 / 2457 [R(int) = 0.0211]
Completeness to θ= 28.34	98.7%
Absorption correction	Empirical
Max. and min. transmission	0.5865 and 0.1946
Refinement method	Full-matrix least-squares on F^2
Data / restraints / parameters	2457 / 0 / 111
Goodness-of-fit on F^2	1.073
Final R indices [I>2σ(I)]	R_1 = 0.0265, wR_2 = 0.0632
R indices (all data)	R_1 = 0.0281, wR_2 = 0.0640
Largest diff. peak and hole	1.475 and -1.362 e.A^{-3}

Appendix

1-Chlorospiro[1λ^3,2-benziodaoxole-3.1'-cyclohexane] (11)

CCDC	771239
Empirical formula	$C_{12}H_{14}ClIO$
Formula weight	336.58
Temperature	100(2) K
Wavelength	0.71073 Å
Crystal system, space group	Monoclinic, $P2_1/c$
Unit cell dimensions	a = 6.0694(16) Å $\alpha = 90°$
	b = 15.369(4) Å $\beta = 101.778(6)°$
	c = 13.039(3) Å $\gamma = 90°$
Volume	1190.7(5) Å^3
Z, Calculated density	4, 1.878 Mg/m^3
Absorption coefficient	2.885 mm^{-1}
F(000)	656
Crystal size	0.41 x 0.20 x 0.04 mm
Data collection	Bruker SMART APEX platform with CCD Detector Graphite monochromator
Detector distance	50 mm
Method; exposure time/frame	omega-scans; t = 1 sec
Solution by	direct methods
Refinement method	full matrix least-squares on F^2, SHELXTL
Theta range for data collection	2.07 to 27.87°
Limiting indices	-7<=h<=7, -20<=k<=20, -17<=l<=17
Reflections collected / unique	11751 / 2832 [R(int) = 0.0428]
Completeness to θ = 27.87	100.0%
Absorption correction	Empirical
Max. and min. transmission	0.8836 and 0.3806
Refinement method	Full-matrix least-squares on F^2
Data / restraints / parameters	2832 / 0 / 136
Goodness-of-fit on F^2	1.034
Final R indices [I>2σ(I)]	R_1 = 0.0290, wR_2 = 0.0647
R indices (all data)	R_1 = 0.0353, wR_2 = 0.0677
Largest diff. peak and hole	1.026 and -0.833 e.A^{-3}

Appendix

1-Chlorospiro[1λ^3,2-benziodaoxole-3.9'-bicyclo[3.3.1]nonane] (12)

CCDC	jcve324
Empirical formula	$C_{15}H_{18}ClIO$
Formula weight	376.64
Temperature	100(2) K
Wavelength	0.71073 Å
Crystal system, space group	Monoclinic, $P2_1/c$
Unit cell dimensions	a = 14.013(3) Å α = 90°
	b = 9.5650(18) Å β = 104.108(4)°
	c = 21.101(4) Å γ = 90°
Volume	2743.0(9) Å^3
Z, Calculated density	8, 1.824 Mg/m^3
Absorption coefficient	2.515 mm^{-1}
F(000)	1488
Crystal size	0.41 x 0.09 x 0.04 mm
Data collection	Bruker SMART APEX platform with CCD Detector Graphite monochromator
Detector distance	50 mm
Method; exposure time/frame	omega-scans; t = 4 sec
Solution by	direct methods
Refinement method	full matrix least-squares on F^2, SHELXTL
Theta range for data collection	1.99 to 27.11 deg.
Limiting indices	-17<=h<=17, -12<=k<=12, -27<=l<=27
Reflections collected / unique	25046 / 6051 [R(int) = 0.0680]
Completeness to θ = 27.11	99.9%
Absorption correction	Empirical
Max. and min. Transmission	0.9061 and 0.4253
Refinement method	Full-matrix least-squares on F^2
Data / restraints / parameters	6051 / 0 / 325
Goodness-of-fit on F^2	1.040
Final R indices [I>2σ(I)]	R_1 = 0.0458, wR_2 = 0.1047
R indices (all data)	R_1 = 0.0590, wR_2 = 0.1106
Largest diff. peak and hole	2.450 and -1.225 e.A^{-3}

Appendix

1-Chloro-3-methoxymethyl-3-methyl-1H,3H-λ^3-dihydro-1,2-benziodoxol (13)

CCDC	771244
Empirical formula	$C_{10}H_{12}ClIO_2$
Formula weight	326.55
Temperature	200(2) K
Wavelength	0.71073 Å
Crystal system, space group	Monoclinic, $P2_1/c$
Unit cell dimensions	a = 10.3999(9) Å $\alpha = 90°$
	b = 14.2086(12) Å $\beta = 94.397(2)°$
	c = 7.5807(6) Å $\gamma = 90°$
Volume	1116.89(16) Å^3
Z, Calculated density	4, 1.942 Mg/m^3
Absorption coefficient	3.078 mm^{-1}
F(000)	632
Crystal size	0.313 x 0.207 x 0.177 mm
Data collection	Bruker SMART APEX platform with CCD Detector Graphite monochromator
Detector distance	50 mm
Method; exposure time/frame	omega-scans; t = 1 sec
Solution by	direct methods
Refinement method	full matrix least-squares on F^2, SHELXTL
Theta range for data collection	1.96 to 27.10°
Limiting indices	-13<=h<=13, -18<=k<=18, -9<=l<=9
Reflections collected / unique	10493 / 2465 [R(int) = 0.0283]
Completeness to θ = 27.10	100.0%
Absorption correction	Empirical
Max. and min. transmission	0.6118 and 0.4458
Refinement method	Full-matrix least-squares on F^2
Data / restraints / parameters	2465 / 0 / 129
Goodness-of-fit on F^2	1.065
Final R indices [I>2σ(I)]	R_1 = 0.0215, wR_2 = 0.0514
R indices (all data),	R_1 = 0.0243, wR_2 = 0.0532
Largest diff. peak and hole	0.576 and -0.396 e.A^{-3}

1-Chloro-3-methyl-3-phenyl-1H,3H-λ^3-dihydro-1,2-benziodoxol (14)

CCDC	771565
Empirical formula	$C_{14}H_{12}ClIO$
Formula weight	358.59
Temperature	100(2) K
Wavelength	0.71073 Å
Crystal system, space group	Monoclinic, $P2_1/c$
Unit cell dimensions	a = 10.7724(10) Å $\alpha = 90°$
	b = 15.9601(14) Å $\beta = 104.101(2)°$
	c = 7.8349(7) Å $\gamma = 90°$
Volume	1306.5(2) Å^3
Z, Calculated density	4, 1.823 Mg/m^3
Absorption coefficient	2.636 mm^{-1}
F(000)	696
Crystal size	0.133 x 0.087 x 0.053 mm
Data collection	Bruker SMART APEX platform with CCD Detector Graphite monochromator
Detector distance	50 mm
Method; exposure time/frame	omega-scans; t = 12 sec
Solution by	direct methods
Refinement method	full matrix least-squares on F^2, SHELXTL
Theta range for data collection	1.95 to 27.88°
Limiting indices	-14<=h<=14, -21<=k<=21, -10<=l<=10
Reflections collected / unique	13055 / 3126 [R(int) = 0.0456]
Completeness to θ = 27.88	99.9%
Absorption correction	None
Refinement method	Full-matrix least-squares on F^2
Data / restraints / parameters	3126 / 0 / 155
Goodness-of-fit on F^2	0.965
Final R indices [I>2σ(I)]	R_1 = 0.0246, wR_2 = 0.0492
R indices (all data)	R_1 = 0.0308, wR_2 = 0.0511
Largest diff. peak and hole	0.932 and -0.426 e.A^{-3}

Appendix

1-Chloro-3-isopropyl-3-phenyl-1,3-dihydro-1,2-benziodoxol (15)

CCDC	771247
Empirical formula	$C_{16}H_{16}ClIO$
Formula weight	386.64
Temperature	200(2) K
Wavelength	0.71073 Å
Crystal system, space group	Orthorhombic, $Pna2_1$
Unit cell dimensions	a = 15.8040(11) Å $\alpha = 90°$
	b = 10.0513(7) Å $\beta = 90°$
	c = 9.4518(7) Å $\gamma = 90°$
Volume	1501.43(18) $Å^3$
Z, Calculated density	4, 1.710 Mg/m^3
Absorption coefficient	2.300 mm^{-1}
F(000)	760
Crystal size	0.58 x 0.45 x 0.36 mm
Data collection	Bruker SMART APEX platform with CCD Detector Graphite monochromator
Detector distance	50 mm
Method; exposure time/frame	omega-scans; t = 1 sec
Solution by	direct methods
Refinement method	full matrix least-squares on F^2, SHELXTL
Theta range for data collection	2.40 to 30.87°
Limiting indices	-22<=h<=21, -14<=k<=13, -13<=l<=12
Reflections collected / unique	16231 / 4370 [R(int) = 0.0283]
Completeness to θ = 30.87	95.0%
Absorption correction	Empirical
Max. and min. Transmission	0.4914 and 0.3488
Refinement method	Full-matrix least-squares on F^2
Data / restraints / parameters	4370 / 1 / 172
Goodness-of-fit on F^2	1.054
Final R indices [I>2σ(I)]	R_1 = 0.0210, wR_2 = 0.0476
R indices (all data)	R_1 = 0.0223, wR_2 = 0.0481
Absolute structure parameter	-0.010(15)
Largest diff. peak and hole	0.434 and -0.601 $e.A^{-3}$

Appendix

1-Chloro-3,3-dimethyl-3a,6-methano-3a,4,5,6,-tetrahydro-1*H*,3*H*-λ^3-ioda-2-oxa-phenalene (16)

CCDC	771243
Empirical formula	$C_{14}H_{16}ClIO$
Formula weight	362.62
Temperature	195(2) K
Wavelength	0.71073 Å
Crystal system, space group	Triclinic, $P\bar{1}$
Unit cell dimensions	a = 8.3813(5) Å, α = 91.4050(10)°
	b = 8.6063(5) Å, β = 106.1170(10)°
	c = 9.7043(6) Å, γ = 103.5340(10)°
Volume	650.80(7) Å^3
Z, Calculated density	2, 1.850 Mg/m^3
Absorption coefficient	2.646 mm^{-1}
F(000)	356
Crystal size	0.53 x 0.51 x 0.40 mm
Data collection	Bruker SMART APEX platform with CCD Detector Graphite monochromator
Detector distance	50 mm
Method; exposure time/frame	omega-scans; t(28) = 1 sec, t(55) = 3 sec
Solution by	direct methods
Refinement method	full matrix least-squares on F^2, SHELXTL
Theta range for data collection	2.19 to 30.48°
Limiting indices	-11<=h<=11, -12<=k<=12, -13<=l<=13
Reflections collected / unique	7405 / 3673 [R(int) = 0.0165]
Completeness to θ = 25.00	99.9%
Absorption correction	Empirical
Max. and min. transmission	0.4167 and 0.3366
Refinement method	Full-matrix least-squares on F^2
Data / restraints / parameters	3673 / 0 / 154
Goodness-of-fit on F^2	1.053
Final R indices [I>2σ(I)]	R_1 = 0.0226, wR_2 = 0.0551
R indices (all data)	R_1 = 0.0238, wR_2 = 0.0558
Largest diff. peak and hole	0.695 and -0.551 e.A^{-3}

Appendix

7-Chloro-5,5-dimethyl-7λ^3-ioda-3-oxa-6λ^5-azatricyclo[6.4.0.0^{2,6}]dodeca-1(8),2(6),9,11-tetraen-6-ylium tetrafluoro-λ^4-borane (20)

CCDC	771242
Empirical formula	$C_{11}H_{12}BClF_4INO$
Formula weight	423.38
Temperature	100(2)
Wavelength	0.71073 Å
Crystal system, space group	Orthorhombic, $P2_12_12_1$
Unit cell dimensions	a = 7.2462(11) Å α = 90°
	b = 11.7348(18) Å β = 90°
	c = 16.806(3) Å γ = 90°
Volume	1429.1(4) Å^3
Z, Calculated density	4, 1.968 Mg/m^3
Absorption coefficient	2.465 mm^{-1}
F(000)	816
Crystal size	0.11 x 0.09 x 0.07 mm
Data collection	Bruker SMART APEX platform with CCD Detector Graphite monochromator
Detector distance	50 mm
Method; exposure time/frame	omega-scans; t_1 = 6 sec, t_2 =12 sec
Solution by	direct methods
Refinement method	full matrix least-squares on F^2, SHELXTL
Theta range for data collection	2.12 to 33.12°
Limiting indices	-11<=h<=11, -18<=k<=18, -25<=l<=25
Reflections collected / unique	52891 / 5429 [R(int) = 0.0913]
Completeness to θ = 33.12	99.9%
Absorption correction	None
Refinement method	Full-matrix least-squares on F^2
Data / restraints / parameters	5429 / 0 / 184
Goodness-of-fit on F^2	1.020
Final R indices [I>2σ(I)]	R_1 = 0.0331, wR_2 = 0.0621
R indices (all data)	R_1 = 0.0371, wR_2 = 0.0632
Absolute structure parameter	0.277(18)
Largest diff. peak and hole	1.877 and -1.434 e.A^{-3}

Appendix

8-Chloro-8λ^3-ioda-7λ^5-azatricyclo-[7.4.0.0^{2,7}]trideca-1(9),2,4,6,10,12-hexaen-7-ylium tetrafluoro-λ^4-borane (21)

CCDC	771238
Empirical formula	$C_{11}H_8NBF_4ClI$
Formula weight	403.34
Temperature	100(2) K
Wavelength	0.71073 Å
Crystal system, space group	Triclinic, $P\bar{1}$
Unit cell dimensions	a = 8.2133(6) Å, α = 95.803(3)°
	b = 8.3060(7) Å, β = 103.756(2)°
	c = 10.5531(8) Å, γ = 110.254(2)°
Volume	642.61(9) Å^3
Z, Calculated density	2, 2.085 Mg/m^3
Absorption coefficient	2.730 mm^{-1}
F(000)	384
Crystal size	0.34 x 0.33 x 0.21 mm
Data collection	Bruker SMART APEX platform with CCD Detector
	Graphite monochromator
Detector distance	50 mm
Method; exposure time/frame	omega-scans; t_1 = 0.5 sec, t_2 = 1.5 sec
Solution by	direct methods
Refinement method	full matrix least-squares on F^2, SHELXTL
Theta range for data collection	2.03 to 33.14°
Limiting indices	-12<=h<=12, -12<=k<=12, -16<=l<=16
Reflections collected / unique	22774 / 4894 [R(int) = 0.0275]
Completeness to θ = 33.14	99.8%
Absorption correction	Empirical
Max. and min. Transmission	0.5940 and 0.4527
Refinement method	Full-matrix least-squares on F^2
Data / restraints / parameters	4894 / 0 / 172
Goodness-of-fit on F^2	1.075
Final R indices [I>2σ(I)]	R_1 = 0.0201, wR_2 = 0.0483
R indices (all data)	R_1 = 0.0209, wR_2 = 0.0487
Largest diff. peak and hole	1.230 and -0.540 e.A^{-3}

1-(Dichloro-λ^3-iodanyl)-2-(1-fluoro-1-methylethyl)benzene (25a)

Empirical formula	$C_9H_{10}Cl_2FI$
Formula weight	334.97
Temperature	100(2) K
Wavelength	0.71073 Å
Crystal system, space group	Orthorhombic, $P2_12_12_1$
Unit cell dimensions	a = 7.7413(12) Å $\alpha = 90°$
	b = 11.4335(18) Å $\beta = 90°$
	c = 12.4612(19) Å $\gamma = 90°$
Volume	1102.9(3) $Å^3$
Z, Calculated density	4, 2.017 Mg/m^3
Absorption coefficient	3.354 mm^{-1}
F(000)	640
Crystal size	0.317 x 0.269 x 0.192 mm
Data collection	Bruker SMART APEX platform with CCD Detector Graphite monochromator
Detector distance	50 mm
Method; exposure time/frame	omega-scans; t = 1 sec
Solution by	direct methods
Refinement method	full matrix least-squares on F^2, SHELXTL
Theta range for data collection	2.42 to 27.88°
Limiting indices	-10<=h<=10, -15<=k<=15, -16<=l<=16
Reflections collected / unique	10538 / 2635 [R(int) = 0.0446]
Completeness to $\theta = 27.88$	100.0%
Absorption correction	Empirical
Max. and min. transmission	0.5653 and 0.4162
Refinement method	Full-matrix least-squares on F^2
Data / restraints / parameters	2635 / 0 / 120
Goodness-of-fit on F^2	1.079
Final R indices [I>2σ(I)]	$R_1 = 0.0275$, $wR_2 = 0.0633$
R indices (all data)	$R_1 = 0.0295$, $wR_2 = 0.0641$
Absolute structure parameter	0.02(3)
Largest diff. peak and hole	1.617 and -1.061 $e.Å^{-3}$

1-(Trifluoromethyl)spiro[1λ^3,2-benziodaoxole-3.1'-cyclohexane] (27)

CCDC	771240
Empirical formula	$C_{13}H_{14}F_3IO$
Formula weight	370.14
Temperature	200(2) K
Wavelength	0.71073 Å
Crystal system, space group	Monoclinic, $C2/c$
Unit cell dimensions	a = 23.244(3) Å α = 90°
	b = 6.5576(9) Å β = 122.141(2)°
	c = 21.344(3) Å γ = 90°
Volume	2754.8(6) Å^3
Z, Calculated density	8, 1.785 Mg/m^3
Absorption coefficient	2.345 mm^{-1}
F(000)	1440
Crystal size	0.294 x 0.289 x 0.073 mm
Data collection	Bruker SMART APEX platform with CCD Detector Graphite monochromator
Detector distance	50 mm
Method; exposure time/frame	omega-scans; t = 3 sec
Solution by	direct methods
Refinement method	full matrix least-squares on F^2, SHELXTL
Theta range for data collection	2.07 to 27.10°
Limiting indices	-29<=h<=29, -8<=k<=8, -27<=l<=27
Reflections collected / unique	12528 / 3054 [R(int) = 0.0352]
Completeness to θ = 27.10	100.0%
Absorption correction	Empirical
Max. and min. transmission	0.8474 and 0.5456
Refinement method	Full-matrix least-squares on F^2
Data / restraints / parameters	3054 / 0 / 163
Goodness-of-fit on F^2	1.096
Final R indices [I>2σ(I)]	R_1 = 0.0355, wR_2 = 0.0848
R indices (all data)	R_1 = 0.0404, wR_2 = 0.0879
Largest diff. peak and hole	1.457 and -0.552 e.A^{-3}

Appendix

1-Trifluoromethyl-3-methyl-3-phenyl-1H,3H-λ^3-dihydro-1,2-benziodoxol (28)

CCDC	771246
Empirical formula	$C_{15}H_{12}F_3IO$
Formula weight	392.15
Temperature	200(2) K
Wavelength	0.71073 Å
Crystal system, space group	Rhombohedral, $R\bar{3}$
Unit cell dimensions	a = 28.0675(7) Å α = 90°
	b = 28.0675(7) Å β = 90°
	c = 9.3727(3) Å γ = 120°
Volume	2.279 mm^{-1}
F(000)	3420
Crystal size	0.83 x 0.23 x 0.17 mm
Data collection	Bruker SMART APEX platform with CCD Detector Graphite monochromator
Detector distance	50 mm
Method; exposure time/frame	omega-scans; t(28) = 1 sec, t(55) = 6 sec
Solution by	direct methods
Refinement method	full matrix least-squares on F^2, SHELXTL
Theta range for data collection	2.33 to 36.34°
Limiting indices	-46<=h<=46, -46<=k<=46, -15<=l<=15
Reflections collected / unique	75525 / 6742 [R(int) = 0.0337]
Completeness to θ = 36.34	97.6%
Absorption correction	Empirical
Max. and min. transmission	0.6980 and 0.2535
Refinement method	Full-matrix least-squares on F^2
Data / restraints / parameters	6742 / 0 / 229
Goodness-of-fit on F^2	1.093
Final R indices [I>2σ(I)]	R_1 = 0.0255, wR_2 = 0.0600
R indices (all data)	R_1 = 0.0280, wR_2 = 0.0610
Largest diff. peak and hole	1.627 and -0.600 e.A^{-3}

Appendix

1-Trifluoromethyl-3,3-dimethyl-3a,6-methano-3a,4,5,6-tetrahydro-1H,3H-λ^3-ioda-2-oxa-phenalene (29)

CCDC	771237
Empirical formula	$C_{15}H_{16}F_3IO$
Formula weight	396.18
Temperature	200(2) K
Wavelength	0.71073 Å
Crystal system, space group	Triclinic, $P\bar{1}$
Unit cell dimensions	a = 8.4585(4) Å α = 96.5060(10)°
	b = 9.4086(4) Å β = 106.6140(10)°
	c = 9.8875(5) Å γ = 105.0460(10)°
Volume	713.07(6) Å^3
Z, Calculated density	2, 1.845 Mg/m^3
Absorption coefficient	2.272 mm^{-1}
F(000)	388
Crystal size	0.80 x 0.50 x 0.43 mm
Data collection	Bruker SMART APEX platform with CCD Detector Graphite monochromator
Detector distance	50 mm
Method; exposure time/frame	omega-scans; t = 2 sec
Solution by	Patterson methods
Refinement method	full matrix least-squares on F^2, SHELXTL
Theta range for data collection	2.20 to 28.27°
Limiting indices	-11<=h<=11, -12<=k<=12, -13<=l<=13
Reflections collected / unique	7367 / 3504 [R(int) = 0.0132]
Completeness to θ = 28.27	99.0%
Absorption correction	Empirical
Max. and min. transmission	0.4445 and 0.2653
Refinement method	Full-matrix least-squares on F^2
Data / restraints / parameters	3504 / 0 / 181
Goodness-of-fit on F^2	1.074
Final R indices [I>2σ(I)]	R$_1$ = 0.0196, wR$_2$ = 0.0500
R indices (all data)	R$_1$ = 0.0205, wR$_2$ = 0.0505
Largest diff. peak and hole	0.485 and -0.703 e.A^{-3}

Appendix

(E)-N-(1-(1H-Benzo[d][1,2,3]triazol-1-yl)ethylidene)trifluoromethanamine (39)

CCDC	792179
Empirical formula	$C_9H_7F_3N_4$
Formula weight	228.19
Temperature	100(2)
Wavelength	0.71073 Å
Crystal system, space group	Monoclinic, $P21/c$
Unit cell dimensions	a = 11.4571(19) Å $\alpha = 90°$
	b = 7.5553(13) Å $\beta = 116.699(3)°$
	c = 12.462(2) Å $\gamma = 90°$
Volume	963.7(3) Å^3
Z, Calculated density	4, 1.573 Mg/m^3
Absorption coefficient	0.141 mm^{-1}
F(000)	464
Crystal size	0.47 x 0.415 x 0.11 mm
Data collection	Bruker SMART APEX platform with CCD Detector Graphite monochromator
Detector distance	50 mm
Method; exposure time/frame	omega-scans; t = 1 sec
Solution by	direct methods
Refinement method	full matrix least-squares on F^2, SHELXTL
Theta range for data collection	3.26 to 27.88°
Limiting indices	-15<=h<=15, -9<=k<=9, -16<=l<=16
Reflections collected / unique	9251 / 2294 [R(int) = 0.0348]
Completeness to θ = 27.88	99.9%
Absorption correction	Empirical
Max. and min. transmission	0.9846 and 0.9365
Refinement method	Full-matrix least-squares on F^2
Data / restraints / parameters	2294 / 0 / 146
Goodness-of-fit on F^2	1.049
Final R indices [I>2σ(I)]	R_1 = 0.0445, wR_2 = 0.1108
R indices (all data)	R_1 = 0.0533, wR_2 = 0.1167
Largest diff. peak and hole	0.392 and -0.198 e.A^{-3}

(E)-N-(1-(2H-Indazol-1-yl)ethylidene)-1,1,1-trifluoromethanamine (54a)

CCDC	792180
Empirical formula	$C_{10}H_8F_3N_3$
Formula weight	227.19
Temperature	100(2)
Wavelength	0.71073 Å
Crystal system, space group	Triclinic, $P\bar{1}$
Unit cell dimensions	a = 5.7075(13) Å α = 82.619(4)°
	b = 7.2373(17) Å b = 86.489(4)°
	c = 11.921(3) Å γ = 78.197(4)°
Volume	477.71(19) Å^3
Z, Calculated density	2, 1.579 Mg/m^3
Absorption coefficient	0.140 mm^{-1}
F(000)	232
Crystal size	0.30 x 0.18 x 0.13 mm
Data collection	Bruker SMART APEX platform with CCD Detector Graphite monochromator
Detector distance	50 mm
Method; exposure time/frame	omega-scans; t = 4 sec
Solution by	direct methods
Refinement method	full matrix least-squares on F^2, SHELXTL
Theta range for data collection	1.72 to 26.02°
Limiting indices	-7<=h<=7, -8<=k<=8, -14<=l<=14
Reflections collected / unique	3966 / 1860 [R(int) = 0.0428]
Completeness to θ = 26.02	99.4
Absorption correction	None
Refinement method	Full-matrix least-squares on F^2
Data / restraints / parameters	1860 / 0 / 146
Goodness-of-fit on F^2	1.077
Final R indices [I>2σ(I)]	R_1 = 0.0506, wR_2 = 0.1071
R indices (all data)	R_1 = 0.0688, wR_2 = 0.1136
Largest diff. peak and hole	0.296 and -0.223 e.A^{-3}

(E)-N-(1-(1H-Indazol-1-yl)ethylidene)-1,1,1-trifluoromethanamine (54b)

CCDC	792181
Empirical formula	$C_{10}H_8F_3N_3$
Formula weight	227.19
Temperature	100(2) K
Wavelength	0.71073 Å
Crystal system, space group	Triclinic, $P\bar{1}$
Unit cell dimensions	a = 6.5664(10) Å α = 98.767(3)°
	b = 8.0025(12) Å β = 99.230(3)°
	c = 9.4686(14) Å γ = 98.872(3)°
Volume	477.16(12) Å^3
Z, Calculated density	2, 1.581 Mg/m^3
Absorption coefficient	0.140 mm^{-1}
F(000)	232
Crystal size	0.32 x 0.17 x 0.06 mm
Data collection	Bruker SMART APEX platform with CCD Detector Graphite monochromator
Detector distance	50 mm
Method; exposure time/frame	omega-scans; t = 4 sec
Solution by	direct methods
Refinement method	full matrix least-squares on F^2, SHELXTL
Theta range for data collection	2.22 to 25.68°
Limiting indices	-7<=h<=7, -9<=k<=9, -11<=l<=11
Reflections collected / unique	3714 / 1795 [R(int) = 0.0203]
Completeness to θ = 25.68	99.6%
Absorption correction	Empirical
Max. and min. transmission	0.9917 and 0.9567
Refinement method	Full-matrix least-squares on F^2
Data / restraints / parameters	1795 / 0 / 145
Goodness-of-fit on F^2	1.089
Final R indices [I>2σ(I)]	R_1 = 0.0448, wR_2 = 0.1058
R indices (all data)	R_1 = 0.0497, wR_2 = 0.1086
Largest diff. peak and hole	0.373 and -0.249 e.A^{-3}

Appendix

(*E*)-*N*-(1-(3,5-Diphenyl-1*H*-pyrazol-1-yl)ethylidene)-1,1,1-trifluoromethanamine (60)

Identification code	792182
Empirical formula	$C_{18}H_{14}F_3N_3$
Formula weight	329.32
Temperature	100(2) K
Wavelength	0.71073 Å
Crystal system, space group	Monoclinic, $C2/c$
Unit cell dimensions	a = 31.118(4) Å $\alpha = 90°$
	b = 5.9041(8) Å $\beta = 125.622(2)°$
	c = 21.482(3) Å $\gamma = 90°$
Volume	3208.3(7) Å^3
Z, Calculated density	8, 1.364 Mg/m^3
Absorption coefficient	0.107 mm^{-1}
F(000)	1360
Crystal size	0.67 x 0.55 x 0.03 mm
Data collection	Bruker SMART APEX platform with CCD Detector Graphite monochromator
Detector distance	50 mm
Method; exposure time/frame	omega-scans; t = 15 sec
Solution by	direct methods
Refinement method	full matrix least-squares on F^2, SHELXTL
Theta range for data collection	1.61 to 25.68°
Limiting indices	-37<=h<=37, -7<=k<=7, -25<=l<=26
Reflections collected / unique	12555 / 3025 [R(int) = 0.0332]
Completeness to θ = 25.68	99.4%
Absorption correction	Empirical
Max. and min. transmission	0.9966 and 0.9315
Refinement method	Full-matrix least-squares on F^2
Data / restraints / parameters	3025 / 0 / 218
Goodness-of-fit on F^2	1.189
Final R indices [I>2σ(I)]	R_1 = 0.0781, wR_2 = 0.2123
R indices (all data)	R_1 = 0.0836, wR_2 = 0.2152
Largest diff. peak and hole	0.486 and -0.314 e.A^{-3}

Appendix

3-(1-Adamantyl)-1-(trifluoromethyl)-1H-pyrazole (79)

CCDC	841860
Empirical formula	$C_{14}H_{17}F_3N_2$
Formula weight	270.30
Temperature	100(2) K
Wavelength	0.71073 Å
Crystal system, space group	Monoclinic, $P2_1/c$
Unit cell dimensions	a = 11.128(2) Å $\alpha = 90°$
	b = 11.807(2) Å $\beta = 119.474(4)°$
	c = 11.174(2) Å $\gamma = 90°$
Volume	1278.1(5) Å^3
Z, Calculated density	4, 1.405 Mg/m^3
Absorption coefficient	0.114 mm^{-1}
F(000)	568
Crystal size	0.30 x 0.19 x 0.10 mm
Data collection	Bruker SMART APEX platform with CCD Detector Graphite monochromator
Detector distance	50 mm
Method; exposure time/frame	omega-scans; t = 4 sec
Solution by	direct methods
Refinement method	full matrix least-squares on F^2, SHELXTL
Theta range for data collection	2.10 to 27.48°
Limiting indices	-14<=h<=14, -15<=k<=15, -14<=l<=14
Reflections collected / unique	12302 / 2932 [R(int) = 0.0617]
Completeness to θ = 27.48	99.9%
Absorption correction	None
Refinement method	Full-matrix least-squares on F^2
Data / restraints / parameters	2932 / 0 / 172
Goodness-of-fit on F^2	0.895
Final R indices [I>2σ(I)]	R_1 = 0.0424, wR_2 = 0.0864
R indices (all data)	R_1 = 0.0670, wR_2 = 0.0937
Largest diff. peak and hole	0.322 and -0.284 e.A^{-3}

1-Trifluoromethyl-5-(2,4,6-trimethylphenyl)pyrazole (80b)

CCDC	841859	
Empirical formula	$C_{13}H_{13}F_3N_2$	
Formula weight	254.25	
Temperature	100(2) K	
Wavelength	0.71073	
Crystal system, space group	Triclinic, $P\bar{1}$	
Unit cell dimensions	a = 7.7188(17) Å	α = 109.075(4)°
	b = 7.9770(18) Å	β = 91.068(4)°
	c = 10.646(2) Å	γ = 91.319(5)°
Volume	619.1(2) Å^3	
Z, Calculated density	2, 1.364 Mg/m^3	
Absorption coefficient	0.113mm^{-1}	
F(000)	264	
Crystal size	0.142 x 0.134 x 0.052 mm	
Data collection	Bruker SMART APEX platform with CCD Detector Graphite monochromator	
Detector distance	50 mm	
Method; exposure time/frame	omega-scans; t = 6 sec	
Solution by	direct methods	
Refinement method	full matrix least-squares on F^2, SHELXTL	
Theta range for data collection	2.02 to 28.26°	
Limiting indices	-10<=h<=10, -10<=k<=10, -14<=l<=14	
Reflections collected / unique	6341 / 3043 [R(int) = 0.0798]	
Completeness to θ = 28.26	99.0%	
Absorption correction	None	
Refinement method	Full-matrix least-squares on F^2	
Data / restraints / parameters	3043 / 0 / 166	
Goodness-of-fit on F^2	0.686	
Final R indices [I>2σ(I)]	R_1 = 0.0456, wR_2 = 0.0693	
R indices (all data)	R_1 = 0.1528, wR_2 = 0.0833	
Largest diff. peak and hole	0.268 and -0.241 e.A^{-3}	

Appendix

Ethyl 3-methyl-1-(trifluoromethyl)-1*H*-pyrazole-4-carboxylate (81a)

CCDC	841861	
Empirical formula	$C_8H_9F_3N_2O_2$	
Formula weight	222.17	
Temperature	100(2) K	
Wavelength	0.71073 Å	
Crystal system, space group	Orthorhombic, *Pbca*	
Unit cell dimensions	a = 12.621(3) Å	$\alpha = 90°$
	b = 10.583(2) Å	$\beta = 90°$
	c = 14.628(3) Å	$\gamma = 90°$
Volume	1953.8(8) Å^3	
Z, Calculated density	8, 1.511 Mg/m^3	
Absorption coefficient	0.146 mm^{-1}	
F(000)	912	
Crystal size	0.27 x 0.13 x 0.03 mm	
Data collection	Bruker SMART APEX platform with CCD Detector Graphite monochromator	
Detector distance	50 mm	
Method; exposure time/frame	omega-scans; t = 10 sec	
Solution by	direct methods	
Refinement method	full matrix least-squares on F^2, SHELXTL	
Theta range for data collection	2.78 to 27.90°	
Limiting indices	-16<=h<=16, -13<=k<=13, -19<=l<=19	
Reflections collected / unique	18461 / 2329 [R(int) = 0.0943]	
Completeness to θ = 27.90	100.0%	
Absorption correction	None	
Refinement method	Full-matrix least-squares on F^2	
Data / restraints / parameters	2329 / 0 / 138	
Goodness-of-fit on F^2	0.888	
Final R indices [I>2σ(I)]	$R_1 = 0.0437$, $wR_2 = 0.0744$	
R indices (all data)	$R_1 = 0.0824$, $wR_2 = 0.0842$	
Largest diff. peak and hole	0.235 and -0.237 e.A^{-3}	

3-Methyl-1-(trifluoromethyl)-5-(2,4,6-trimethylphenyl)-1H-pyrazole (85a)

CCDC	841862
Empirical formula	$C_{14}H_{15}F_3N_2$
Formula weight	268.28
Temperature	100(2)
Wavelength	0.71073 Å
Crystal system, space group	Triclinic, $P\bar{1}$
Unit cell dimensions	a = 8.500(3) Å, α = 67.349(10)°
	b = 8.772(4) Å, β = 89.438(7)°
	c = 11.330(4) Å, γ = 61.026(7)°
Volume	665.6(5) Å^3
Z, Calculated density	2, 1.339 Mg/m^3
Absorption coefficient	0.109 mm^{-1}
F(000)	280
Crystal size	0.19 x 0.10 x 0.07 mm
Data collection	Bruker SMART APEX platform with CCD Detector Graphite monochromator
Detector distance	50 mm
Method; exposure time/frame	omega-scans; t = 15 sec
Solution by	direct methods
Refinement method	full matrix least-squares on F^2, SHELXTL
Theta range for data collection	2.00 to 26.73°
Limiting indices	-10<=h<=10, -11<=k<=11, -14<=l<=14
Reflections collected / unique	6142 / 2805 [R(int) = 0.0549]
Completeness to θ = 26.73	99.1%
Absorption correction	None
Refinement method	Full-matrix least-squares on F^2
Data / restraints / parameters	2805 / 12 / 204
Goodness-of-fit on F^2	1.102
Final R indices [I>2σ(I)]	R_1 = 0.0650, wR_2 = 0.1828
R indices (all data)	R_1 = 0.0997, wR_2 = 0.1952
Largest diff. peak and hole	0.350 and -0.274 e.A^{-3}

4,5-Diphenyl-1-(trifluoromethyl)-1H-1,2,3-triazole (89a)

CCDC	843429
Empirical formula	$C_{15}H_{10}F_3N_3$
Formula weight	289.
Temperature	100(2)
Wavelength	0.71073 Å
Crystal system, space group	Orthorhombic, $P2_12_12_1$
Unit cell dimensions	a = 5.8907(15) Å α = 90°
	b = 12.633(3) Å β = 90°
	c = 17.711(5) Å γ = 90°
Volume	1318.0(6) Å^3
Z, Calculated density	4, 1.458 Mg/m^3
Absorption coefficient	0.119 mm^{-1}
F(000)	592
Crystal size	0.16 x 0.09 x 0.04mm
Data collection	Bruker SMART APEX platform with CCD Detector Graphite monochromator
Detector distance	50 mm
Method; exposure time/frame	omega-scans; t = 40 sec
Solution by	direct methods
Refinement method	full matrix least-squares on F^2, SHELXTL
Theta range for data collection	1.98 to 28.47°
Limiting indices	-7<=h<=7, -16<=k<=16, -23<=l<=23
Reflections collected / unique	13651 / 3273 [R(int) = 0.1156]
Completeness to θ = 28.47	99.2%
Absorption correction	None
Refinement method	Full-matrix least-squares on F^2
Data / restraints / parameters	3273 / 0 / 190
Goodness-of-fit on F^2	0.944
Final R indices [I>2σ(I)]	R_1 = 0.0670, wR_2 = 0.0818
R indices (all data)	R_1 = 0.1016, wR_2 = 0.0904
Largest diff. peak and hole	0.259 and -0.239 e.A^{-3}

Ethyl 6,7,8-trifluoro-4-(trifluoromethoxy)quinoline-3-carboxylate (94)

Empirical formula	$C_{13}H_7F_6NO_3$
Formula weight	339.20
Temperature	100(2) K
Wavelength	0.71073 Å
Crystal system, space group	Triclinic, $P\bar{1}$
Unit cell dimensions	a = 4.847(2) Å, α = 69.910(7)°
	b = 11.158(5) Å, β = 84.242(7)°
	c = 12.548(5) Å, γ = 78.603(7)°
Volume	624.4(5) Å^3
Z, Calculated density	2, 1.804 Mg/m^3
Absorption coefficient	0.18 mm^{-1}
F(000)	340
Crystal size	0.30 x 0.03 x 0.01 mm
Data collection	Bruker APEX2 with Bruker APEX-II CCD Detector Graphite monochromator
Detector distance	50 mm
Method; exposure time/frame	omega-scans; t = 60 sec
Solution by	direct methods
Refinement method	full matrix least-squares on F^2, SHELXTL
Theta range for data collection	1.73 to 26.02°
Limiting indices	-5<=h<=5, -13<=k<=13, -15<=l<=15
Reflections collected / unique	8359 / 2447 [R(int) = 0.1534]
Completeness to θ = 26.02	100.0%
Absorption correction	None
Refinement method	Full-matrix least-squares on F^2
Data / restraints / parameters	2447 / 0 / 209
Goodness-of-fit on F^2	0.788
Final R indices [I>2σ(I)]	R_1 = 0.0499, wR_2 = 0.0840
R indices (all data)	R_1 = 0.1641, wR_2 = 0.1111
Largest diff. peak and hole	0.282 and -0.362 e.Å^{-3}

i want morebooks!

Buy your books fast and straightforward online - at one of world's fastest growing online book stores! Environmentally sound due to Print-on-Demand technologies.

Buy your books online at
www.get-morebooks.com

Kaufen Sie Ihre Bücher schnell und unkompliziert online – auf einer der am schnellsten wachsenden Buchhandelsplattformen weltweit! Dank Print-On-Demand umwelt- und ressourcenschonend produziert.

Bücher schneller online kaufen
www.morebooks.de

VDM Verlagsservicegesellschaft mbH
Heinrich-Böcking-Str. 6-8
D - 66121 Saarbrücken

Telefon: +49 681 3720 174
Telefax: +49 681 3720 1749

info@vdm-vsg.de
www.vdm-vsg.de

Printed by Books on Demand GmbH, Norderstedt / Germany